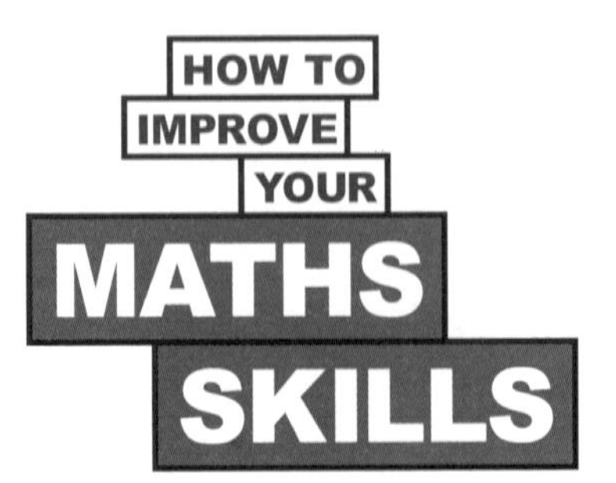
HOW TO
IMPROVE
YOUR
MATHS
SKILLS

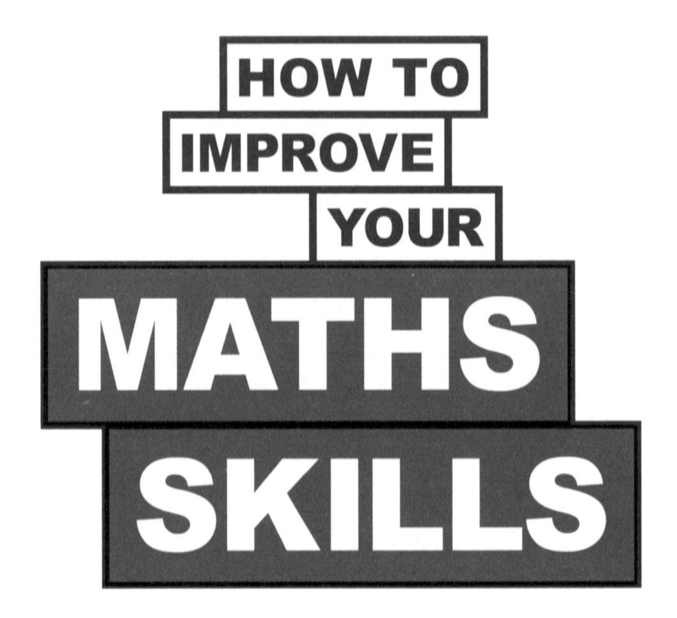

STEVE LAKIN

ALWAYS LEARNING PEARSON

Harlow, England • London • New York • Boston • San Francisco • Toronto • Sydney
Auckland • Singapore • Hong Kong • Tokyo • Seoul • Taipei • New Delhi
Cape Town • Sao Paulo • Mexico City • Madrid • Amsterdam • Munich • Paris • Milan

Pearson Education Limited
Edinburgh Gate
Harlow
Essex CM20 2JE
England

and Associated Companies throughout the world

Visit us on the World Wide Web at:
www.personed.co.uk

First published 2010
Rejacketed edition published 2011

ISBN: 978-0-273-77002-2

British Library Cataloguing-in-Publication Data
A catalague record for this book is available from the British Library

Library of Congress Cataloging-in-Pulication Data
A catalog record for this book is available from the Library of Congress

10 9 8 7 6 5 4 3 2 1
14 13 12 11

Typeset in 10/13pt Din Regular by 3
Printed and bound in Great Britain by Ashford Colour Press Ltd, Gosport, Hampshire

Contents

Preface and acknowledgements

Maths – aarrgghhhhhh!

If that's your first thought, don't worry, you are not alone.

However, you actually use maths all the time in your daily life without even thinking about it – every time you get a shopping bill, do banking, divide up food, and in thousands of other ways, you are actually doing maths.

As a student, no matter what course you are doing, there will be a need for some mathematics. The aim of this book is to introduce you gently to the basic topics needed, to start to take away the fears you may have of mathematics, and to provide you with the basic skills in number, algebra, data presentation, and other topics, that you will need during your studies.

Hopefully, the book will build your confidence and show you that this sort of mathematics is in everyday use in your life, and that you can succeed with it.

This book would not have been possible without the support of my colleagues in the Mathematics Division at the University of Glamorgan, to whom I am very grateful. I would also like to express my sincere thanks to my former PhD supervisor Rick Thomas at the University of Leicester, whose encouragement, inspiration and support will never be forgotten.

I would also like to thank Pearson Education for supporting this publication, in particular Steve Temblett and Katy Robinson.

Finally, I would like to acknowledge my family whose constant support has always been there – special thanks to my wife Dasha for absolutely everything.

Steve Lakin

How to use this book

This book is intended to be a supportive text to give you confidence in the sort of maths that you will encounter both in your studies and in your daily life.

In the book we cover the basic ideas in arithmetic, algebra and data, and also some other topics that are likely to be useful for you. It's not meant to be a comprehensive guide to the whole of mathematics, just something to give you the basics and make you see the relevance of maths to whatever you are doing.

It is intended to be read as a whole – so working through from the first chapter to the last – but you might find it useful just to 'dip in' to the sections that you need. Use it in whatever way is helpful to you – just remember that some sections may refer to topics covered in previous sections.

Each chapter talks you through a particular topic, with examples illustrating the idea. Included in each chapter are a couple of 'smart tips' which hopefully you can take on board and use to relate the mathematics to your study and your daily life. Always try to keep in mind that mathematics is not meant to be a foreign language, it's meant to be useful – also the amount of times you actually need it and use it every day.

At the end of each chapter is a set of exercises for you to try. Don't despair if you find some of them hard – the intention is to provide a mix of questions for each topic.

Try to avoid using a calculator unless you have to. I've provided you with a times table which I'd prefer you to use rather than a calculator – of course for some questions you will have to resort to a calculator, but try as much as you can to avoid its use – you don't always have a calculator to hand in your daily life, and if you have less reliance on it, you'll gain confidence.

You can find solutions at the back of the book, but these are only the final answers. There is a companion website to this book, which you can link to from the Pearson Smarter Student site at **www.smarterstudyskills.com**.

On this website, you can find worked solutions to all the exercises, together with further questions (including some related to particular fields of study), contact details, and more. If you really don't understand a particular answer, go to the website and look at the full worked solution, and see if that makes things clearer.

The book is meant to help you. Maths isn't natural to a lot of people but it's my hope that this book will make you see its relevance to your daily life and to your studies. You can succeed with maths, and hopefully this book will help you do just that!

Times tables

I know that learning times tables isn't particularly fun, and I know that you can resort to a calculator. But you'll gain much more confidence if you can tackle maths on a daily basis (when you probably don't have a calculator to hand) and perform calculations in your head. This can help you when shopping, for example, so you can keep track of what you are spending.

It would be good if you knew the multiplications ('times tables') up to 12. To try to build your confidence and wean you off the calculator, I am providing you with the following table, which you can use to read off any multiplication of numbers from 1 to 12.

	1	**2**	**3**	**4**	**5**	**6**	**7**	**8**	**9**	**10**	**11**	**12**
1	1	2	3	4	5	6	7	8	9	10	11	12
2	2	4	6	8	10	12	14	16	18	20	22	24
3	3	6	9	12	15	18	21	24	27	30	33	36
4	4	8	12	16	20	24	28	32	36	40	44	48
5	5	10	15	20	25	30	35	40	45	50	55	60
6	6	12	18	24	30	36	42	48	54	60	66	72
7	7	14	21	28	35	42	49	56	63	70	77	84
8	8	16	24	32	40	48	56	64	72	80	88	96
9	9	18	27	36	45	54	63	72	81	90	99	108
10	10	20	30	40	50	60	70	80	90	100	110	120
11	11	22	33	44	55	66	77	88	99	110	121	132
12	12	24	36	48	60	72	84	96	108	120	132	144

When doing the questions in the book, try not to use a calculator unless you have to – but you can look at this table. To use it, find the first number you want to multiply on the left, find the second number along the top, and find the intersection of them. For example, for 5×7, find 5 on the left, 7 along the top, and the square where they 'cross each other' is 35, which is the answer.

You don't have to use this table – but maybe it will help you get the 'times tables' into your head rather than being reliant on a calculator.

ARITHMETIC

1 Basic arithmetic and the 'BODMAS' rule

Getting used to basic arithmetic and simple calculations, and making sure you perform them correctly!

You need to have a grasp of the sort of basic numerical calculations that you will encounter on a day-to-day basis both in your studies and in your personal life. This chapter reviews the basic rules of arithmetic and introduces you to the very important BODMAS rule to ensure that you perform your calculations correctly.

Key topics

→ Addition, subtraction, multiplication and division
→ Quotients and remainders
→ The BODMAS rule

Key terms

arithmetic addition subtraction multiplication division brackets quotient remainder BODMAS

Arithmetic is the word used to describe the basic calculations that you can do with numbers. There are four basic *operations* (don't worry about the word 'operation', it just means 'something that does something to numbers') in arithmetic – addition, subtraction, multiplication and division.

Note that you really should be able to do these sort of calculations in your head, even if you have to use your fingers or use a piece of paper, so try not to use a calculator at all for this chapter. Learning times tables is a drag but, if you can learn them, it can save you a lot of time when you have to do calculations in daily life or exams, when you don't have a calculator to hand.

→ Addition

An example of addition is combining two lengths together to produce a longer length. If you work for six days and then you work for another three days, how many days have you worked in total? Easy enough, nine days. This can be written as $6 + 3 = 9$ (said as '6 plus 3 equals 9').

Note that this addition (or *sum*) is unchanged if you do the three days first, and then go on to do the six days (you've still worked for nine days), so that $3 + 6 = 6 + 3$ (both sums equal 9). The fact that it doesn't matter which way round you add things up is quite an important property – technically the word for this is *commutative,* but you can probably get through life without knowing that. If you have a whole list of numbers added together, you can change the order in which they are added, and you still get the same answer. For example,

$3 + 7 + 22 + 2 = 22 + 7 + 3 + 2$ (check for yourself that both give 34)

Addition crops up in finance very often. For example, if you have £40 in your bank account and you deposit (pay in) another £30, then your balance is now £40 + £30 = £70.

→ Subtraction

Subtraction is in some sense the 'opposite' to addition. Using a similar example to above, suppose I have nine days to work and I've already worked three days. How many more days do I have to work? The answer is clearly six days. This is the same as saying 'what do I have to add to three to get nine?'. This can be written as $9 - 3 = 6$ (said as '9 minus 3 equals 6'). The word *subtract* is also used instead of *minus*.

Note, importantly, that the order of subtraction *cannot* be swapped like we did for addition. For example, $8 - 3$ is not the same as $3 - 8$ (in fact $3 - 8$ is a *negative number* -5, you will see this in Chapter 2).

As another example, again with your bank account, if you have £70 in your account and you withdraw (take out) £50 then your balance is now £70 − £50 = £20.

→ Multiplication

An example of multiplication is combining two lengths together to produce an area. Suppose I have a room which is three metres by six metres – how much carpet do I need to furnish it? The answer is 18 square metres, written as $3 \times 6 = 18$ (said as '3 multiplied by 6 equals 18' – the word *times* is also used instead of 'multiplied'.

The order of multiplication *can* be swapped – note for example that $3 \times 6 = 6 \times 3$ (both sides are 18), a useful aid when learning times tables. So again, if you have a long list of things to be multiplied together, you can multiply them in any order you like – check for yourself that $3 \times 10 \times 2 \times 4 = 4 \times 2 \times 10 \times 3$ for example (both equal 240).

A very common use of multiplication is when shopping. For example, if books cost £3 each and you buy 7 books, what is the total cost? The total cost is 7 lots of £3, that is 7×3 which is 21, and so the total cost is £21.

→ Division

Division is in some sense the 'opposite' of multiplication. Suppose we know that the area of a room is 18 square metres and one side is 6 metres, then how long is the other side?

This is the same as asking the question 'What do I have to multiply 6 by, in order to produce 18?' The answer is 3 and so our other side is 3 metres long. This is written as $18 \div 6 = 3$ (read as '18 divided by 6 equals 3').

The order of division *cannot* be swapped: $15 \div 5$ is not the same as $5 \div 15$. In fact, the second one cannot be done using whole numbers and needs fractions, which you will see in Chapter 3.

Again, as another example, suppose you buy 5 books and the total cost was £20. How much was each book? The answer is $20 \div 5 = 4$ and so each book is £4.

Start to realise how much maths plays a part in your daily life by realising how you are using maths in everyday situations, such as shopping, banking, calculating travel times, splitting restaurant bills, and so many other things. Every time you do anything involving numbers, think about what you are doing, and whether you are using addition, subtraction, multiplication or division. If you start to realise how much you actually use maths every day without even thinking about it, you will increase your confidence and realise the point of it and why we need it.

→ Quotients and remainders

Often when dividing, you cannot give a whole number as the answer. Although $20 \div 5$ can easily be done to get 4, what happens when you do $20 \div 6$? You can't divide 6 exactly into 20, so we get a bit stuck. We'll see how to deal with this when we do fractions, but note that we can also approach this another way.

Informally $20 \div 6$ means 'how many 6s go into 20'? If you think about it, then you'll see that 6 goes exactly into 18 (which is close to 20) three times, and you have 2 left over. Mathematically you can see this as $20 = 3 \times 6 + 2$ but you might think of it more informally as something along the lines of 'How many times does 6 divide into 20? Answer: 3 times with 2 left over'.

The number of times it divides is called the *quotient* and the number 'left over' is called the *remainder*. So, in our example above, the quotient when you divide 20 by 6 is 3 ('6 goes into 20 three times'), and the remainder is 2 ('2 left over').

→ The 'BODMAS' rule

Suppose you have a sum like $3 + 4 \times 5$. There's a problem here – which operation is done first, addition or multiplication? If you do the addition first, then you do $3 + 4 = 7$ and so you work out $7 \times 5 = 35$. If you do the multiplication first, then you do $4 \times 5 = 20$ and so you work out $3 + 20 = 23$. Which of these is right?

The rule is that **multiplication is always done before addition**. So the second way is right: $3 + 4 \times 5 = 3 + 20 = 23$

It does make sense to do things this way, especially when you think about shopping. If you buy 2 books for £3 and 4 CDs for £5 then what is the total cost? In your head, you can probably work out that this is £26. Now mathematically, writing it all on one line, the total cost is $2 \times 3 + 4 \times 5$. Doing the multiplication before the addition, this is the same as $6 + 20 = 26$ and so the cost is £26, which is what it should be and what you'd expect. (If you just worked from left to right, working each step out one after the other, you would get £50, which is clearly wrong, and you wouldn't be happy about it if that was your bill!). You do this all the time when you are shopping, so you have been using the fact that multiplication is done before addition without really thinking about it!

What if you really do want to do the addition first? Say bottles of juice are £2 each. You buy three, and like them so much you buy another five. How much have you spent in total? Well, in total you have bought $3 + 5 = 8$ bottles, at £2 each, and so you have spent $8 \times 2 =$ £16. How would you write this on one line? You can't write $3 + 5 \times 2$ because this means do the multiplication first, which would give $3 + 10 = 13$ which is the wrong answer.

If you really want to force addition to be done first, then you must put the addition *in brackets*. Something in brackets is always worked out first, even before multiplication. So writing this as $(3 + 5) \times 2$ gives (doing the brackets first) $8 \times 2 = 16$ which is the right answer.

This convention (brackets always done first, and multiplication comes before addition) is part of a general convention which tells you which things to do first and in what order to do them. This states that you do things in this order:

1. Brackets
2. Orders (which you don't need to know about for now)
3. Divisions and Multiplications
4. Additions and Subtractions

So always do the brackets first, then any divisions and multiplications, then do any additions and subtractions last.

For example, what is $3 + 2 \times (5 + 2) + 1 - 3 + 2$?

First work out the brackets: $5 + 2 = 7$, so we get $3 + 2 \times 7 + 1 - 3 + 2$.

Next is the multiplication: $2 \times 7 = 14$ so we get $3 + 14 + 1 - 3 + 2$.

Now we have additions and subtractions, so just work this out from left to right: this computes to 17.

This convention is known as BODMAS (standing for Brackets, Orders, Division, Multiplication, Addition, Subtraction). You *must* learn this and always remember to apply it in calculations.

The BODMAS rule

B Always work out any *brackets* first.

O Ignore this for now, it stands for *orders*, but we include it just because the acronym BODMAS is easier to remember than BDMAS!

D, **M** } After brackets, work out any *divisions* and *multiplications*

A, **S** } Finally work out the *additions* and *subtractions*, going from left to right.

smart tip

When you go shopping, try to keep a rough track of how much you are spending overall as you buy various quantities of different things, and see if you can predict your final bill. As well as helping you budget, this will help you remember why the BODMAS rule is so important.

Summary

You use basic arithmetical calculations all the time without even realising it. Every time you go shopping you are basically applying the BODMAS rule to work out your bill, every time you use your bank account you are adding or subtracting a number from your account balance. In your studies, no matter what course you are doing, you will have to perform basic calculations on a regular basis. If you can become comfortable now in manipulating numbers and doing these calculations quickly and correctly, you will save yourself a lot of time and greatly boost your confidence!

Exercises

Try to complete the following exercises. As with every chapter of this book, each set of questions is preceded by an example. Solutions are at the back of the book to check your answer, but obviously don't look at them until you have tried the questions – there would be no point to that! If you don't get the right answer, go back and see if you can work out where you went wrong. Remember that further material, including fully worked solutions and further questions (including some relevant to your field of study), is available on the book's website.

Don't use a calculator – get confident with doing calculations yourself (though use pen and paper if you like).

1 Show that it doesn't matter in what order you do addition and multiplication by calculating the following and showing that you get the same answer.

Example: $2 + 3 + 4$ and $3 + 4 + 2$

Solution: $2 + 3 + 4 = 9$ and $3 + 4 + 2 = 9$ so the answer is the same.

(a) $3 + 5 + 2$ and $2 + 3 + 5$ (b) $5 + 6 + 15 + 4$ and $15 + 5 + 6 + 4$

(c) 2×3 and 3×2 (d) $2 \times 5 \times 6$ and $6 \times 2 \times 5$

2 What are the quotients and remainders in the following divisions?

Example: $17 \div 5$

Solution: $17 = 3 \times 5 + 2$ and so the quotient is 3 and the remainder is 2 (you might say '5 goes into 17 three times, with remainder 2').

(a) $22 \div 6$ (b) $14 \div 3$ (c) $33 \div 7$ (d) $24 \div 6$

3 Work out the following (remember the BODMAS rule!)

Example: $2 + 3 \times 5$

Solution: $2 + 3 \times 5 = 2 + 15$ (doing the multiplication first) $= 17$.

(a) $3 + 4 \times 2$ (b) $7 + 2 \times 4$ (c) $15 - 5 \times 2$

(d) $4 \times 5 + 2$ (e) $3 \times 6 + 2 \times 5$ (f) $4 \times 6 - 3 \times 5$

(g) $24 \div 6 + 2$ (h) $4 + 6 \div 2$ (i) $4 + 3 \times 5 - 2 \times 4$

4 Work out the following (remember the BODMAS rule, brackets first!)

Example: $4 \times (2 + 3)$

Solution: $4 \times (2 + 3) = 4 \times 5$ (doing the brackets first) $= 20$.

(a) $6 \times (5 + 2)$ (b) $4 \times (8 - 6)$ (c) $(3 + 4) \times (2 + 5)$

(d) $(6 - 2) \times (9 - 2)$ (e) $6 + (2 + 3) \times 5$ (f) $(5 + 2 \times 3) + 4 \times (3 + 4)$

5 Answer the following questions related to shopping.

Example: If you buy 3 books at £5 each, and 4 posters at £3 each, what is the total bill? What is the change from £30?

Solution: The calculation is $3 \times 5 + 4 \times 3 = 15 + 12 = 27$, so your bill is £27. The change is £30 − £27 = £3.

(a) What is the total bill if you download 3 movies at £5 each, 4 videos at £3 each, and 2 albums at £2 each?

(b) If you buy 4 bottles of drink at £2 each and 3 ready meals at £3 each, what is the change from a £20 note?

(c) If you buy 3 packets of crisps at 40p each, 4 cans of drink at 60p each, and 5 chocolate bars at 25p each, what is your total bill, and what is the change from a £5 note?

(d) Two supermarkets, MegaDeals and BuyNow, both sell frozen pizza and frozen chips. MegaDeals charges £3 for pizza, and £2 for chips. BuyNow charges £5 for pizza and £4 for chips, but is running a special offer of 'buy one, get one free' on both products. Which supermarket is cheaper if you need four of each product?

2 Negative numbers

Understanding negative numbers and dealing with them accurately

Negative numbers appear in daily life quite often, for example with temperatures or (more unfortunately) in finance. It is important that you understand the purpose of negative numbers and how to manipulate them in the same way that you do 'normal' numbers.

Key topics

→ Negative numbers
→ Addition, subtraction, multiplication and division with negative numbers

Key terms
number line positive number negative number

'Normal' counting starts at 1 and proceeds to 2, 3, 4, 5 etc – that is, all the whole numbers. Numbers are often used to represent increases in quantities, things that go up. However there is often a need to represent decreases in quantities, things that go down. This is where we need negative numbers.

→ Negative numbers on the number line

You can write the 'normal' numbers as a *number line*, which is simply a horizontal line with the numbers labelled on it. You can include the negative numbers by extending the number line to the left, to get something like the following line (you can imagine how it goes on for ever in each direction).

The numbers to the left of 0 (smaller than 0) are called *negative numbers*. 'Normal numbers' bigger than 0 (to the right) are called *positive numbers*.

One way in which you already use negative numbers is in temperature. When it's 0°C (zero degrees Celsius) and it gets a bit colder, it drops to −1°C, −2°C etc. Think of the number line above as a thermometer and it should start to make sense.

smart tip

> Get used to negative numbers by finding out the temperature of various cities around the world, particularly those in their winter season, and working out which cities are colder (so for example a city with temperature −5°C is colder than a city with temperature −2°C, and even colder than a city with temperature 3°C).

Bear in mind that negative numbers basically represent the 'opposite' to positive numbers. Some examples of where negative numbers might be used in the real world:

- If a positive number means a bank account is in credit, a negative number means it is in debit (overdrawn).
- If a positive number means to increase something, a negative number means to decrease it.
- If a positive number means a rise, a negative number means a fall.

Often, to avoid confusion in expressions, negative numbers are written with brackets around them, like (−3) and (−5) and so on.

→ Addition and subtraction with negative numbers

To add and subtract when negative numbers are involved, look again at the 'number line' above. Follow the basic rule that:

- **When you add a positive number, you move to the right, and when you subtract a positive number, you move to the left.**
- **When you add a negative number, you move to the left, and when you subtract a negative number, you move to the right.**

The second rule does make some sort of sense, since negatives are in some sense the 'opposite' to positives, so instead of going right you go left, and vice versa.

Some examples to illustrate this:

- What is $3 + 5$? This is just normal addition and you can do it in your head, but following our rules, this is adding a positive number (5) and so go to 3 on the number line and move 5 places to the right. The answer is 8.
- What is $(-2) + 3$? This is adding a positive number (3) and so go to -2 on the number line and move 3 places to the right. The answer is 1.
- What is $5 - 8$? This is subtracting a positive number (8) and so go to 5 on the number line and move 8 places to the left. The answer is -3.
- What is $(-3) + (-4)$? This is adding a negative number (-4) and so go to -3 on the number line and move 4 places to the left. The answer is -7.
- What is $2 - (-5)$? This is subtracting a negative number (-5) and so go to 2 on the number line and move 5 places to the right. The answer is 7.

The last one here is the hardest one to remember. Subtracting a negative number means go *up* and increase (as opposed to subtracting a positive number, which means go down, or decrease).

Try not to get too hung up on this. Think about temperatures – adding a positive number means it gets warmer and subtracting means it gets colder.

Example

- The temperature is currently 2 degrees. It gets five degrees colder. What is the temperature now? The temperature is now $2 - 5$ which, if you check on the number line, is -3.
- Now it gets another 5 degrees colder. The temperature is now $(-3) - 5 = -8$ degrees.
- Now it gets 2 degrees warmer. The temperature is now $-8 + 2 = -6$ degrees.

→ Multiplication and division with negative numbers

Multiplication and division when negative numbers are involved is very similar to normal multiplication and division but with one important fact to remember.

Basically, you multiply or divide the numbers as if they were positive, and then make sure your answer is either positive or negative depending on the following:

- **If both the numbers are positive, or both the numbers are negative, the answer must be positive.**
- **Otherwise, if one is positive and one is negative, then the answer must be negative.**

Examples

- $3 \times 4 = 12$ (since both are positive, the answer is positive)
- $3 \times (-4) = -12$ (since one is positive, and one negative, the answer is negative)
- $(-3) \times 4 = -12$ (since one is positive, and one negative, the answer is negative)
- $(-3) \times (-4) = 12$ (since both are negative, the answer is positive)

The one that causes the most problems is the last one: remember that a negative number multiplied by a negative number is a *positive* number.

Similarly with division:

- $15 \div 5 = 3$ (since both are positive, the answer is positive)
- $15 \div (-5) = -3$ (since one is positive, and one negative, the answer is negative)
- $(-15) \div 5 = -3$ (since one is positive, and one negative, the answer is negative)
- $(-15) \div (-5) = 3$ (since both are negative, the answer is positive)

Again, remember that a negative number divided by a negative number is a *positive* number.

For many of the topics you will encounter in mathematics, it is easy to create your own examples to test yourself. For example for this topic, just make up some numbers yourself (either positive or negative) and try to add, subtract, multiply or divide them. You can check your answers on a calculator, or by asking someone else.

→ Another example of negative numbers

As well as temperature, negative numbers also arise commonly in finance.

- Suppose you have £50 in your bank account, and you pay in a cheque for £20. Your balance is now £70 since $50 + 20 = 70$.
- Next, your rent of £200 is taken from your account. Your balance is now $70 - 200 = -130$. Your balance is negative – this means you are overdrawn by £130 .
- Next, your electricity payment of £40 is taken. Your balance is now $(-130) - 40 = -170$. Now you are £170 overdrawn.
- The bank penalises your unauthorised overdraft by doubling your overdraft amount. So your balance is now $2 \times (-170) = -340$. You are £340 overdrawn.
- Thankfully it's payday. You pay in your wages of £800. Now your balance is $-340 + 800 =$ £460

These sort of calculations are going on every day in millions of personal banking accounts – you can see it's important to get them right!

Summary

Don't be frightened of negative numbers. You do use them naturally when you think about temperature and finance, for example. They have to exist, otherwise what would be the temperature if it gets colder than zero degrees? What would be the balance in your account if you go overdrawn? Learn the rules given to you and get used to using them, and you'll start to gain the confidence to realise that using negative numbers is not really any harder than using the 'normal' numbers you're more comfortable with!

Exercises

Don't use a calculator – get confident with doing calculations yourself. It may help you to draw a number line on a piece of paper and use that to help you.

1 Work out the following.

Example: $(-2) + 6$

Solution: 4 (go to -2 on the number line and move 6 places to the right).

(a) $5 - 7$ (b) $6 + (-3)$ (c) $(-3) + 5$ (d) $(-12) + 8$

(e) $4 + (-5)$ (f) $(-3) - 5$ (g) $(-2) + (-8)$ (h) $(-9) - (-7)$

(i) $4 + (-4)$ (j) $(-2) - (-2)$ (k) $3 + (-2) - 4$ (l) $1 - 3 - 5 - (-2)$

2 Work out the following.

Example: $3 \times (-5)$

Solution: -15 (the answer must be negative since it is a positive multiplied by a negative).

(a) $2 \times (-4)$ (b) $(-4) \times 6$ (c) $(-2) \times (-8)$

(d) $3 \times (-9)$ (e) $(-10) \times (-10)$ (f) $18 \div (-3)$

(g) $(-21) \div 7$ (h) $(-12) \div (-6)$ (i) $(-10) \div (-10)$

3 Work out the following (remember the BODMAS rule!).

Example: $4 - 2 \times (3 + 2)$

Solution: -6, since this is the same as $4 - 2 \times 5$ (doing the brackets first), which is $4 - 10$ (doing the multiplication next), which is -6.

(a) $5 - 2 \times (4 + 3)$ (b) $5 - 2 \times (8 - 5)$ (c) $3 + 4 \times ((-2) - 3)$

(d) $(2 - 5) \times (3 - 7)$ (e) $4 - 4 \div (6 - 3 - 5)$ (f) $(-3) + (6 + (-2)) \div (-2)$

4 Answer the following questions relating to temperature.

Example: The temperature is falling by 3 degrees every hour. If it is currently 5°C, what will be the temperature in 1 hour? What about 2 hours? 6 hours?

Solution: In 1 hour it will be $5 - 3 = 2$°C. In 2 hours it will be another 3 degrees colder, so $2 - 3 = -1$°C. In 6 hours, it will be $5 - 3 \times 6$ (dropping by 3 degrees every hour for 6 hours) $= -13$°C.

(a) The pressure is dropping in a pressurised container. As a consequence, the temperature is dropping by 5°C every hour. The temperature in the container started at 20°C.

(i) What is the temperature after 2 hours?

(ii) What is the temperature after 6 hours?

(iii) When is the temperature 0°C?

(b) The room temperature is currently 30°C and it is rising by 2°C every hour. A new fan system will take 2 hours to take effect, but then will reduce the temperature by 3°C every hour until the temperature reaches the 22°C standard. How long (from now) will it take for the room to reach the standard 22°C?

3 Fractions

Understanding what fractions mean and being able to manipulate them correctly

Lots of times in life, the things we deal with aren't our normal nice 'whole numbers' but involve parts of a number, or dividing a number into pieces. This is where fractions come in. You will learn what a fraction is and what it represents, and how to add, subtract, multiply and divide them.

Key topics

→ Fractions
→ Mixed and top-heavy fractions
→ Addition, subtraction, multiplication and division of fractions

Key terms

fraction numerator denominator lowest terms mixed top-heavy

We are used to dealing with whole numbers in our daily lives. The temperature is 25 degrees, it is 10 miles to work, and so on. But not everything is dealt with in whole numbers. What if the temperature is above 25 but less than 26? Or the distance is between 1 mile and 2 miles? We will look at decimal numbers later, but for this topic we will look at fractions as expressing a certain number of 'bits' of something.

You do use fractions without really thinking about it – say you are travelling 20 miles, after 10 miles you can think 'we're halfway there'.

Essentially, a fraction is just a division. We saw before that whilst you could do something like $8 \div 2$ to get the answer 4, you weren't able to do something like $2 \div 8$ using whole numbers. Let us see how fractions can be useful and how to manipulate them. There are a fair few rules to learn here, but with practice they will become second nature.

→ Basic fractions

For example, the following pictures (imagine them as cakes or pizza, maybe?) represent 1 bit out of 2 (usually called a half), 1 bit out of 3 (usually called a third) and 1 bit out of 4 (usually called a quarter).

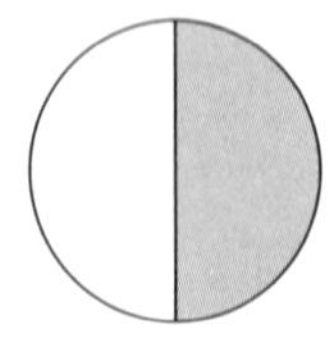 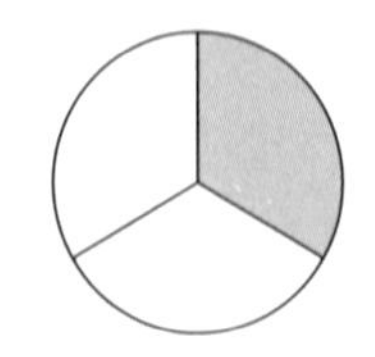 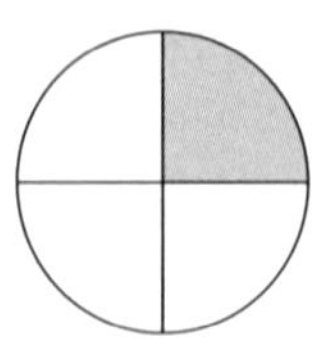

The first picture is '1 bit out of 2' and is written as $\frac{1}{2}$.

The second picture is '1 bit out of 3' and is written as $\frac{1}{3}$.

The third picture is '1 bit out of 4' and is written as $\frac{1}{4}$.

> **smart tip**
>
> In fractions, the number on the top is called the *numerator* and the number on the bottom is called the *denominator*. These are long words, but learn them because they will appear in any books etc – it's fine when you think for yourself, though, if you just think of 'top' and 'bottom'!

So, what do the following pictures represent?

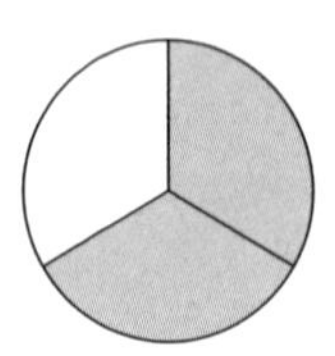 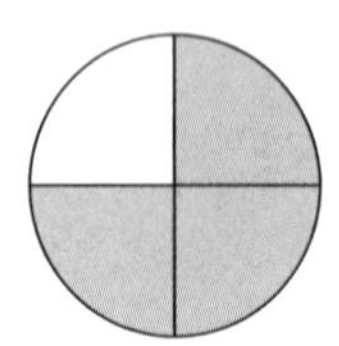

The first is 2 bits out of 3, and is written as $\frac{2}{3}$.

The second is 3 bits out of 4, and is written as $\frac{3}{4}$.

Now look at the following pictures:

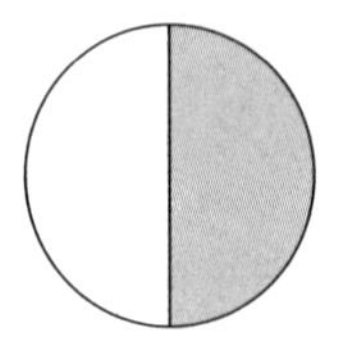 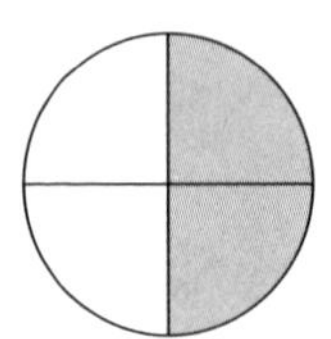 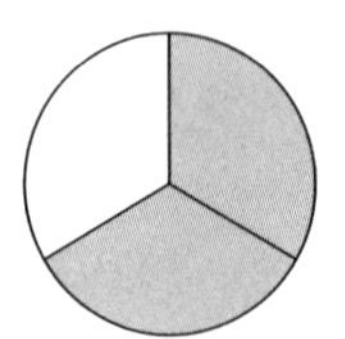 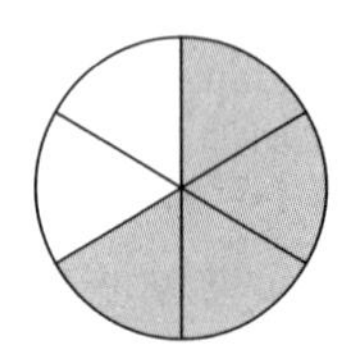

Notice that the first two pictures are 'the same' – they represent exactly the same thing. It's like cutting your pizza into half and giving your friend a half, or cutting it into quarters and giving them two bits – it's all the same amount of pizza. The same applies with the second two pictures.

So $\frac{1}{2}$ is equivalent to ('the same as') $\frac{2}{4}$, and $\frac{2}{3}$ is equivalent to ('the same as') $\frac{4}{6}$.

Also look at the following pictures:

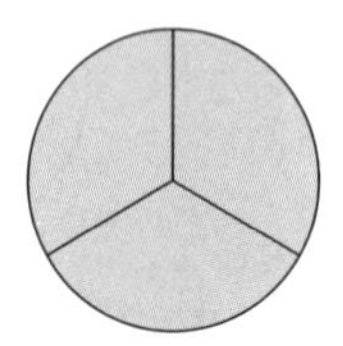 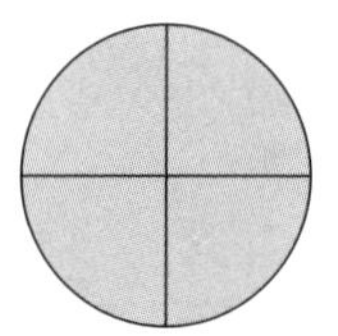

Here, we have 3 bits out of 3 and it makes the whole thing (1 pizza): the same thing with 4 bits out of 4. So 3 bits out of 3 makes 1, and so does 4 bits out of 4. So we have $\frac{3}{3} = 1$ and $\frac{4}{4} = 1$. You can probably guess that this works for any numbers.

Similarly, if you have something like $\frac{3}{1}$ then this is 3 lots of a whole thing, which is just 3 things, and so $\frac{3}{1}$ is just equal to 3. The same goes for other numbers such as $\frac{6}{1} = 6$.

Finally, look at the following picture. This represents a whole thing, all in one picture. What are we representing here?

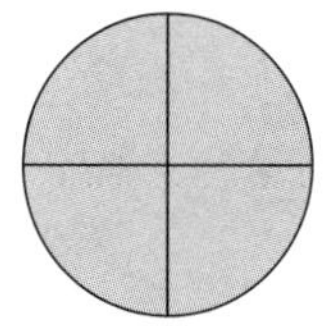 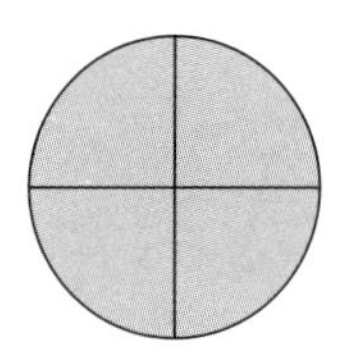 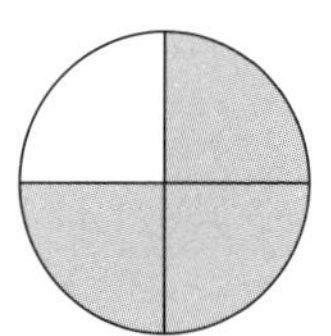

There are two ways to look at it. Overall, we are counting bits of 4 (quarters). We have 11 bits in total, so we can write this as $\frac{11}{4}$. Alternatively, we have two 'wholes' and then three quarters. So you could write it as '2 and 3 quarters' which is written as $2\frac{3}{4}$.

Generally (you'll see why shortly), it is easier to work with the first way, so writing $\frac{11}{4}$.

The first type of fraction is often referred to as a *top-heavy* (or *improper*) fraction and the second type as a *mixed* fraction.

→ Converting between mixed and top-heavy fractions

Usually we will work with top-heavy fractions. If you want to convert fractions from one form to another, there are simple steps to follow.

To convert a top-heavy fraction to a mixed fraction

- Divide the denominator (bottom) into the numerator (top).
- The quotient (how many times the denominator divides into the numerator) is the whole number.
- The fraction is now the remainder, over the original denominator.

It's easier to see this by an example – let's convert $\frac{14}{3}$ into a mixed fraction.

- The numerator is 14 and the denominator 3. Dividing 3 into 14 gives quotient 4 and remainder 2 ('3 goes into 14, 4 times with remainder 2').
- So the whole number is the quotient 4, and the fraction is $\frac{2}{3}$ (the remainder 2, over the original denominator 3)
- Hence our mixed fraction is $4\frac{2}{3}$.
- So $\frac{14}{3}$ is the same as $4\frac{2}{3}$.

Converting from a mixed fraction to a top-heavy fraction

Again there are simple rules to convert the other way round.

- Multiply the whole number by the denominator (bottom).
- Add this to the numerator (top).
- Write the new fraction as this answer, over the original denominator (bottom).

Using the same example as before, $4\frac{2}{3}$:

- Multiply the whole number, 4, by the denominator, 3, to get 12.
- Add this to the numerator 2, to get 14.
- So the new fraction is this answer (14) over the original denominator (3) and we get the answer we expected of $\frac{14}{3}$.

→ Cancelling fractions to their lowest terms

We noted above that $\frac{1}{2}$ is equivalent to $\frac{2}{4}$ and $\frac{2}{3}$ is equivalent to $\frac{4}{6}$. We would always like to make our fractions as simple as possible – this is called cancelling them to their *lowest terms*.

The general rule is this: if a number divides into both the top and bottom of the fraction, then you can 'cancel' it – which means divide by it. To reduce a fraction to its lowest terms, you 'cancel' as much as you can.

So for example consider $\frac{6}{9}$.

3 divides into both the top and the bottom. 6 divided by 3 gives 2, and 9 divided by 3 gives 3. So this fraction is equivalent to $\frac{2}{3}$. This is clear when you look at the following picture, which shows $\frac{2}{3}$ and $\frac{6}{9}$ are identical.

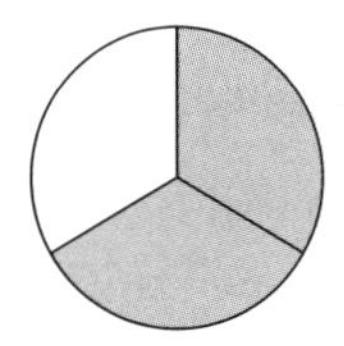
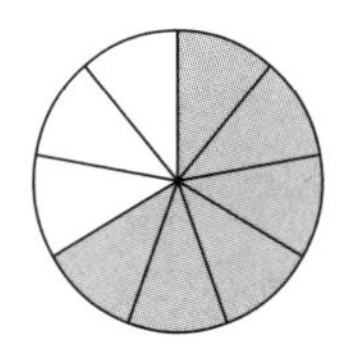

So, to cancel a fraction, you divide anything you can into both the top and the bottom. So as another example, consider $\frac{24}{30}$. The number 6 divides into both the top and the bottom: 6 divides into 24 four times and 6 divides into 30 five times. So this is the same as $\frac{4}{5}$.

smart tip

Get used to the idea of fractions every time you cut a cake or pizza or anything you divide into pieces – if you normally cut it in half, next time cut it into quarters and realise that two quarters is the same as one half. Alternatively, draw circles for yourself and divide them up into pieces. Bringing the mathematical ideas into your daily life will make them seem real and keep them in your mind, building your confidence.

Fractions are just numbers really, so we should be able to add, subtract, multiply and divide them. What, for example, is 'a half times a half' or 'a quarter minus two thirds'? You will be given a set of precise rules to learn how to manipulate fractions; hopefully you will understand why and how they work but, at the very least, make sure you learn the rules!

→ Multiplying fractions

What is a half of a quarter? Think of it in terms of a picture like we have done, and you'll see that a half of a quarter is an eighth (half of a bit of four, is a bit of eight).

There is a basic rule that you can learn for multiplying two fractions together.

Multiply the numerators together, and multiply the denominators together, and then cancel the fraction to its lowest terms.

So, for example, work out $\frac{2}{3} \times \frac{3}{4}$.

Multiply the numerators (top bits) together, and multiply the denominators (bottom bits) together. You get

$$\frac{2 \times 3}{3 \times 4} = \frac{6}{12} = \frac{1}{2}$$

(cancelling the factor of 6).

So the answer is $\frac{1}{2}$.

However, in practice, we can often make life a bit easier for ourselves by effectively 'doing the cancelling first'. All this means is that when we write the multiplication, we can first of all cancel anything that is a factor both somewhere on the top, and somewhere on the bottom.

For example, what is $\frac{9}{8} \times \frac{16}{15}$?

Using the rule from above, this is the same as $\frac{9 \times 16}{8 \times 15} = \frac{144}{120}$ which cancels to $\frac{6}{5}$ (cancelling 24). The problem here is that we had to involve quite big numbers in our multiplying. As an alternative, we can 'do the cancelling first'. Have a look at our fraction. Note that the number 3 divides into both the top and the bottom – it divides into 9 on the top and 15 on the bottom. So we can cancel this to get $\frac{3}{8} \times \frac{16}{5}$ (note that the 16 and the 8 stay the same). Next, notice that 8 divides into 16 on the top and 8 on the bottom, so cancelling this 8, we get $\frac{3}{1} \times \frac{2}{5}$. Now nothing more cancels, and we get $\frac{3 \times 2}{1 \times 5} = \frac{6}{5}$ just like before. The advantage of doing the cancelling 'first' like this is that we don't have such big numbers to deal with.

→ Dividing fractions

The opposite of multiplying is dividing, so dividing by a half should be the same as multiplying by its opposite. What is the opposite of a half? Well, a half is one divided by two, so its opposite is two divided by one, which is two.

Don't worry if this isn't clear to you at first, just learn the rule you are given below. The rule for dividing is as follows:

Turn the second fraction upside-down and then multiply the two fractions together.

So, for example, work out $\frac{2}{3} \div \frac{3}{4}$.

Turn the second one upside down and multiply, so you need to work out

$$\frac{2}{3} \times \frac{4}{3} = \frac{2 \times 4}{3 \times 3} = \frac{8}{9}$$

In this case there is no cancelling to do, but of course do it if necessary.

→ Adding fractions

Adding fractions is somewhat trickier. What is a half plus a quarter? You'd like to think that you could just add up the top bits and add up the bottom bits, but sadly you can't.

You might hope the answer is $\frac{1}{2} + \frac{1}{4} = \frac{1+1}{2+4} = \frac{2}{6} = \frac{1}{3}$. Sorry, this isn't right!

It should be obvious that this can't be right – a third is smaller than a half so it can't be the answer if I'm adding a quarter to the half!

A nice way to see what the answer is, is to convert the two fractions so they both count out of the same bits. In this case, we have a half which is the same as 2 bits out of 4, together with our quarter which is 1 bit out of 4:

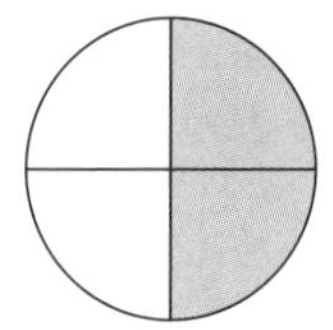

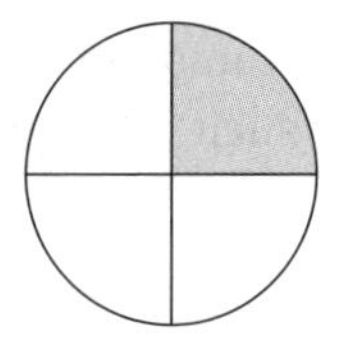

It's now clear that we have two quarters and one quarter, so three quarters in total. To write this mathematically:

$$\frac{1}{2}+\frac{1}{4}=\frac{2}{4}+\frac{1}{4}=\frac{3}{4}$$

The trick in adding is to 'make the numbers on the bottom the same' and then you can do the adding up easily. I'll give you the rule again, which you can just learn until this becomes clear. But first of all you need to know what the *lowest common multiple* (LCM) is.

The lowest common multiple of two numbers is the smallest number that both of them divide into. So for example, take 3 and 4. The smallest number they both divide into is 12. So 12 is the lowest common multiple. Now try 8 and 12. The smallest number they both divide into is 24. So 24 is the lowest common multiple.

Here is your rule for adding up fractions (this one is a bit longer):

- **Work out the LCM (lowest common multiple) of the two denominators**
- **Multiply top and bottom of both fractions so the denominator is the LCM**
- **Add up the numerators and leave the denominators the same**

This needs an example, so here goes:

Add up $\frac{1}{5}+\frac{2}{3}$

Follow the steps *exactly*.

- The two denominators are 5 and 3. The LCM (the smallest number they both divide into) of these two numbers is 15.
- Convert both fractions to have denominator 15. You should find that $\frac{1}{5}$ is the same as $\frac{3}{15}$ (multiplying top and bottom by 3), and $\frac{2}{3}$ is the same as $\frac{10}{15}$ (multiplying top and bottom by 5). This means we have

$$\frac{3}{15}+\frac{10}{15}$$

- Add up the numerators (top bits) and leave the denominator (bottom) as it is

$$\frac{3}{15}+\frac{10}{15}=\frac{3+10}{15}=\frac{13}{15}$$

→ Subtracting fractions

The rule for subtracting is very similar to adding up. You can see (draw a similar picture) that if you take a quarter away from a half, you should get a quarter left over. The rule is:

- **Work out the LCM (lowest common multiple) of the two bottom numbers.**
- **Multiply top and bottom of both fractions so the bottom number is the LCM.**
- **Subtract the top bits and leave the bottom the same.**

The only difference to adding, is that you take away. So, as an example:

Work out $\frac{3}{8} - \frac{1}{3}$

Follow the steps *exactly*.

- The two denominators are 8 and 3. The LCM (the smallest number they both divide into) of these two numbers is 24.
- Convert both fractions to have denominator 24. Similarly to before, this means we have

$$\frac{9}{24} - \frac{8}{24}$$

- Subtract the top bits and leave the bottom as it is

$$\frac{9}{24} - \frac{8}{24} = \frac{9-8}{24} = \frac{1}{24}$$

→ Manipulating mixed fractions

How do you add, subtract, multiply, and divide mixed fractions? It's simple – just convert them to top-heavy fractions and then apply the basic rules above. This is why it's usually better to leave fractions as top-heavy rather than write them as mixed – if you need to manipulate them again later you'd have to convert them back.

Example

What is $2\frac{2}{3} \times 1\frac{1}{4}$?

Converting to top-heavy fractions, this is the same as

$\frac{8}{3} \times \frac{5}{4} = \frac{40}{12} = \frac{10}{3}$ (which you might then write as $3\frac{1}{3}$ if you really want to).

A topic like this has a lot of rules to learn. The best way to learn rules like this is to practise using them over and over again until you become so comfortable with them that you don't even realise you learnt them! Do as many examples as you can – there are examples of fractions at the end of this chapter, on the website, and in numerous other websites and books too, and you can make your own examples. Practise, practise, practise, and you won't have to learn rules, applying them will become natural!

Summary

Fractions are necessary in arithmetic – you can't do much dividing without them. If you view them simply as 'so many bits of something' they seem more natural and you can then apply the rules to manipulate them. It is important that you understand them as they do crop up in all sorts of situations – basically anywhere where you have to divide numbers, so in any situation where you have to split up some quantity of things!

The rules are boring to learn but hopefully you have some sort of grasp as to why the rules are as they are – it does make it easier to learn rules if they make some sort of sense!

Exercises

Don't use a calculator – try to do this yourself just with pen and paper. Give all answers as fractions in their lowest terms.

Give answers as top-heavy (rather than mixed) fractions where appropriate.

1 Just to make sure you have remembered which is which, write down the numerator and denominator of the following fractions.

Example: $\frac{4}{5}$

Solution: Numerator is 4, denominator is 5.

(a) $\frac{3}{7}$ (b) $\frac{8}{9}$ (c) $\frac{11}{3}$

2 Cancel the following fractions to their lowest terms:

Example: $\frac{12}{16}$

Solution: 4 divides into both the top and the bottom, so dividing both by 4, this is the same as $\frac{3}{4}$ ($12 \div 4 = 3$ and $16 \div 4 = 4$). You cannot divide both top and bottom by anything else, so this is now in its lowest terms.

(a) $\frac{3}{6}$ (b) $\frac{6}{24}$ (c) $\frac{18}{27}$

(d) $\frac{28}{63}$ (e) $\frac{8}{100}$ (f) $\frac{7}{15}$

In the following questions 3) to 6), for those involving mixed fractions, remember to convert them into top-heavy fractions first. Also remember to give your answers as fractions in their lowest terms.

3 Work out the following multiplications. For the ones involving mixed fractions, remember to convert them into top-heavy fractions first. Remember to give your answers as fractions in their lowest terms.

Example: $\frac{2}{5} \times \frac{3}{4}$

Solution: Either multiply the numerators and the denominators together to get $\frac{2 \times 3}{5 \times 4} = \frac{6}{20}$ which cancels to $\frac{3}{10}$ (dividing top and bottom by 2), or start by cancelling the 2 from the top and bottom to get $\frac{1}{5} \times \frac{3}{2} = \frac{3}{10}$.

(a) $\frac{1}{2} \times \frac{3}{4}$ (b) $\frac{2}{3} \times \frac{3}{5}$ (c) $\frac{5}{9} \times \frac{3}{2}$

(d) $\frac{5}{6} \times \frac{1}{5}$ (e) $1\frac{1}{2} \times \frac{2}{3}$ (f) $2\frac{1}{3} \times 1\frac{3}{4}$

4 Work out the following divisions.

Example: $\frac{3}{5} \div \frac{5}{6}$

Solution: Turn the second fraction upside-down to get $\frac{3}{5} \times \frac{6}{5}$ and then multiply (as in the question above) to get $\frac{18}{25}$ which does not cancel any further.

(a) $\frac{1}{3} \div \frac{1}{2}$ (b) $\frac{2}{3} \div \frac{3}{4}$ (c) $\frac{2}{5} \div \frac{2}{3}$

(d) $\frac{3}{4} \div \frac{2}{3}$ (e) $\frac{2}{9} \div 1\frac{1}{3}$ (f) $1\frac{2}{3} \div 2\frac{1}{4}$

5 Work out the following additions.

Example: $\frac{2}{3} + \frac{1}{5}$

Solution: The lowest common multiple is 15. So this is the same as $\frac{10}{15} + \frac{3}{15} = \frac{13}{15}$ (multiplying top and bottom of the first fraction by 5, and multiplying top and bottom of the second fraction by 3)

(a) $\frac{1}{3} + \frac{1}{4}$ (b) $\frac{2}{3} + \frac{3}{7}$ (c) $\frac{4}{3} + \frac{5}{9}$

(d) $\frac{2}{5} + \frac{3}{4}$ (e) $\frac{5}{6} + 1\frac{2}{3}$ (f) $1\frac{1}{4} + 2\frac{3}{4}$

6 Work out the following subtractions.

Example: $\frac{4}{5} - \frac{3}{4}$

Solution: The lowest common multiple is 20. So this is the same as $\frac{16}{20} - \frac{15}{20}$ $= \frac{1}{20}$ (multiplying top and bottom of the first fraction by 4, and multiplying top and bottom of the second fraction by 5).

(a) $\frac{1}{3} - \frac{1}{4}$ (b) $\frac{2}{3} - \frac{4}{7}$ (c) $\frac{11}{4} - \frac{9}{4}$

(d) $\frac{9}{11} - \frac{2}{7}$ (e) $1\frac{4}{5} - \frac{2}{3}$ (f) $2\frac{1}{4} - 1\frac{1}{3}$

7 Answer the following 'real-life' questions using fractions.

Example: A jug contains $1\frac{2}{3}$ litres of juice. It is used to fill 3 glasses of $\frac{1}{4}$ litre each. How much juice is left, as a fraction?

Solution: You start with $1\frac{2}{3}$ litres and subtract three lots of $\frac{1}{4}$ litres, so now have $1\frac{2}{3} - \frac{3}{4} = \frac{5}{3} - \frac{3}{4} = \frac{20}{12} - \frac{9}{12} = \frac{11}{12}$ litres.

(a) As in the example above, you then pour another 2 glasses of $\frac{1}{3}$ litre each. How much juice is left now? Can you fill any more glasses?

(b) A computer disk holds $4\frac{1}{2}$ GB. If you put 5 files of size $\frac{2}{3}$ GB each, and 5 files of size $\frac{1}{6}$ GB each onto this disk, how much space is left as a fraction?

4 Percentages, ratios and proportions

Using percentages and ratios to describe and compare values

Percentages are commonly used in finance, the media, in education, and in many walks of life, because they provide an understandable way to present figures that people seem to understand naturally. You are quite used to getting exam results, or seeing election polls on the news, as percentages.
In this topic we will show that percentages are really just fractions, we will show how to create them and use them to compare different situations.

Key topics

→ Percentages
→ Ratios and proportions

Key terms
percentage ratio proportion

Having done fractions in the last topic, percentages are a natural next step, because a percentage is no more than a fraction with denominator 100, as you will see.

→ Percentages

We saw in the last topic that fractions can be 'equivalent', so, for example:

$$\frac{1}{2} = \frac{2}{4} = \frac{3}{6} \ldots \text{etc}$$

To express a fraction as a percentage means to change the fraction to an equivalent one, so that the bottom (called the *denominator*) is 100.

So, for example, what is a half as a percentage? Well, the equivalent fraction to a half, is $\frac{50}{100}$. You can show this is the same – you could (although I don't recommend it, it will take ages) draw a picture like we did before and show that 50 bits out of 100 is exactly half of them, or better, you could *cancel* a factor like we did before.

Remember, we said that if a number divides both the top and bottom, you can 'cancel' the fraction by dividing by that number:

50 divides both the top and the bottom of $\frac{50}{100}$, and so dividing by 50, we get exactly $\frac{1}{2}$. Hence $\frac{1}{2}$ is equivalent to (the same as) $\frac{50}{100}$.

We normally use the symbol % to represent a percentage. This symbol is a sort of 'shorthand' for dividing by 100. To write one of these percentages, instead of writing it as a fraction with bottom part (denominator) 100, we write the top (numerator) followed by the % sign. So in this case, we would write 50%.

As another example, what is a quarter as a percentage? Well, a quarter is $\frac{1}{4}$ which is equivalent to $\frac{25}{100}$ (check this for yourself). And so a quarter, as a percentage, is 25%.

Why do we use percentages? It's simply quite natural for people to rate things out of a hundred, you have probably seen this with exams and so on – your mark is often given as a percentage and you are told that you need, say, 40% to pass.

To work out the corresponding percentage for a fraction, convert it to something out of 100, and that is your percentage. Here are some examples:

Example 1

In a test, you need to get more than 40% to pass. You score 9 out of 20. Have you passed? $\frac{9}{20}$ is equivalent to $\frac{45}{100}$ (multiply top and bottom by 5) and so you got 45%.

So you passed!

Example 2

You are told you scored 75% in a test. The test was out of 20. What was your score out of 20?

To answer this, you need to convert the percentage to a fraction out of 20. Well, $\frac{75}{100}$ is equivalent to $\frac{15}{20}$ (divide top and bottom by 5) and so you got 15 out of 20.

Remember: percentages are just an equivalent fraction, written out of 100.

It's worth remembering some common percentages such as 25% for a quarter, 50% for a half, and 75% for three-quarters.

Fractions like a third get more complicated – there's no real equivalent way to write a third out of 100. We'll see more of this when we do decimals in the next topic, but we will ignore this problem for now.

Now you know that percentages are just fractions, you can do calculations such as working out 25% of 16. This is the same as

$$\frac{25}{100} \times 16 = \frac{1}{4} \times 16 = \frac{16}{4} = 4$$

and so 25% of 16 is 4. (*If that doesn't quite make sense to you done like that, note that 16 is the same as* $\frac{16}{1}$ *and so you perform the multiplication* $\frac{25}{100} \times \frac{16}{1} = \frac{1}{4} \times \frac{16}{1} = \frac{16}{4} = \frac{4}{1} = 4$.)

smart tip

Get used to percentages by converting all of your marks that you get in your tests etc. into percentages. Also see how many times you see percentages used in the media or in banking for example (inflation and interest rates, election polls, sporting data). The more you realise that this sort of mathematics is all around you, the more confident you will become.

→ Ratios

Suppose in a class there are 20 boys and 10 girls.

You can say straight away that 'there are twice as many boys as girls', or another way of looking at it is to say 'for every two boys, there is one girl'. This is an example of a *ratio* – for every two boys there is one girl, which can be written in the form 2:1 (which says, for every 2 boys, there is 1 girl).

You could write the ratio as 4:2 (for every 4 boys, there are 2 girls) or 10:5 (for every 10 boys, there are 5 girls) or 20:10 (for every 20 boys there are 10 girls) and so on. This is very similar to what we did with equivalent fractions – it's easier to write 2:1 but all these ratios are equivalent.

The rule you need to follow to work out a ratio is this:

- **Write down the two numbers.**
- **'Cancel' them by dividing by any number that divides into them both.**

By doing this, you make the ratio as simple as possible.

Example 1

In another class there are 12 boys and 16 girls. What is the ratio of boys to girls?

Start by writing down the two numbers – the number of boys followed by the number of girls. So you write down 12:16

Now look at these numbers 12 and 16. Notice that 4 divides into them both:

$4 \times 3 = 12$ and $4 \times 4 = 16$, so you can 'cancel' the factor of 4.

So 12:16 is equivalent to 3:4, and hence the ratio of boys to girls is 3:4

It gets a bit trickier the other way round, when you are given a ratio and you have to work out the numbers. Here's another example:

Example 2

In another class, there are 15 pupils and the ratio of boys to girls is 2:1 (so twice as many boys as girls). How many boys and how many girls are there?

If you think about this for a while, you'll probably come up with the answer of 10 boys and 5 girls.

The ratio 2:1 is equivalent to the ratio 10:5. What we have done is convert the ratio 2:1 to an equivalent form: just like with fractions, we multiplied both numbers by some other number, in this case 5.

$2 \times 5 = 10$ and $1 \times 5 = 5$ so 2:1 is equivalent to 10:5

Note, that this is just like doing $\frac{2}{1} = \frac{10}{5}$

The crucial point about why we changed it, is that 10 and 5 add up to make 15, the number of pupils.

This does seem to involve a lot of guesswork. How did we know to multiply by 5, did we try 3 (giving 6:2) and 4 (giving 8:4) and 6 (giving

12:6) and so on and realise they didn't work to make 15 students? We could do this, but think for a little while.

The statement 'the ratio of boys to girls is 2:1' means that for every two boys, there is one girl. So I could divide my class of 15 into groups of 3, each with 2 boys and 1 girl.

How many groups are there? There are 15 pupils and each group has 3 people in it so there are $\frac{15}{3} = 5$ groups. That's where the 5 came from. The ratio told us we could split the group into threes, and we divided the total number 15 by 3 to work out how many groups, and hence what to multiply the ratio by.

So here is the rule to answer questions like this:

- **Write down the ratio.**
- **Convert this to an equivalent ratio where the two numbers add up to the total that you are looking for. To do this:**
  - **Add up the two numbers in your ratio,**
  - **Then divide the total you are looking for by this number,**
  - **Then multiply each term of the ratio by this answer above.**

Let's illustrate this with another example.

Example 3

Suppose there are 20 students and the ratio of boys to girls is 2:3. How many boys and girls are there?

Follow the steps exactly:

- Write down the ratio 2:3
- Add up the two numbers in the ratio – you get $2 + 3 = 5$
- Divide the number of students (20) by this number, so $\frac{20}{5} = 4$

Multiply each term of the ratio by this answer 4, so 2:3 becomes 8:12

There are 8 boys and 12 girls.

Of course, this method works for any situation, not just boys and girls in a classroom!

→ Proportions

The word 'proportion' tends to be used quite a lot in the news – 'a large proportion of citizens will be affected by these changes ...' but this word is not really much different from a percentage or ratio.

Example 1

Say you have two classes – one has 6 boys and 6 girls and the other has 12 boys and 8 girls? Which class would you say has the highest 'proportion' of girls?

In the first class, $\frac{6}{12} = 50\%$ of the students are girls.

In the second class, $\frac{8}{20} = 40\%$ of the students are girls.

So although the second class has more girls in it, they are outnumbered by the boys. So, the 'proportion' of girls in the second class is less – in the second class only 40% are girls, but in the first class it's 50%.

Example 2

Suppose we have a class of 4 boys and 6 girls, and another class of 8 boys and 12 girls. Which has the higher proportion of girls?

The percentage of girls in the first class is $\frac{6}{10} = 60\%$

The percentage of girls in the second class is $\frac{12}{20} = 60\%$

So they both have the same proportion of girls in them (60%) – you might say that the two classes are *proportionate.*

Proportion is a useful way to measure the spread of things – if one school has 500 boys and 50 girls, and another school has 40 boys and 40 girls, it is clear that even though the first school has more girls in total, the spread in the second school is far more even.

smart tip

Practise with ratios and proportions by doing your own investigations in daily situations. For example, work out the ratio of men to women in your class, or work out the ratio of 'home-grown' to 'overseas' players in your favourite football team, for example. Use the mathematical concepts you are learning when you can, and in something relevant to you, and you will enjoy them more and find them easier!

Summary

Percentages are used because they come naturally to us, for some reason we as humans find it easier to measure things out of 100 than anything else. But a percentage really is nothing more than a fraction with denominator 100. When you see percentages, remember that they are just fractions!

Exercises

Don't use a calculator – try to do these exercises yourself using only pen and paper.

1 Express the following fractions as percentages.

Example: $\frac{3}{50}$

Solution: $\frac{3}{50} = \frac{6}{100} = 6\%$

(a) $\frac{7}{10}$ (b) $\frac{3}{5}$ (c) $\frac{43}{50}$ (d) $\frac{7}{25}$

2 Give the following test scores as percentages.

Example: 9 out of 10

Solution: As a fraction, 9 out of 10 $= \frac{9}{10} = \frac{90}{100} = 90\%$

(a) 6 out of 10 (b) 11 out of 20

(c) 21 out of 30 (d) 150 out of 200

3 Express the following percentages as fractions in their lowest terms.

Example: 70%

Solution: $70\% = \frac{70}{100} = \frac{7}{10}$ cancelling 10.

(a) 60% (b) 5% (c) 96% (d) 33%

4 Calculate the following.

Example: 50% of 80

Example: 50% of 80 $= \frac{50}{100} \times 80 = \frac{1}{2} \times 80 = \frac{80}{2} = 40.$

(a) 50% of 60 (b) 20% of 80 (c) 25% of 20 (d) 90% of 50

5 Work out the following ratios.

Example: In a box of chocolates, there are 12 plain chocolates and 8 milk chocolates. What is the ratio of plain chocolates to milk chocolates?

Solution: The ratio is 12:8 = 3:2 cancelling a factor of 4.

(a) I have 10 CDs and 5 DVDs in a box. What is the ratio of CDs to DVDs in the box?

(b) In a box of apples, there are 21 red apples and 28 green apples. What is the ratio of red apples to green apples?

6 Work out the answers to the following ratio-based questions.

Example: I have 20 pens, some of which are black and the others blue. The ratio of black pens to blue pens is 3:2. How many black pens are there, and how many blue pens?

Solution: Following the rules that I gave you earlier, add up the numbers in the ratio to get 3 + 2 = 5. Then divide the total number 20 by this number, to get 20 ÷ 5 = 4. Finally multiply the ratio by this number 4, to get 12:8. There are 12 black pens and 8 blue pens.

(a) I have 12 playing cards in total, some are black and some are red. The black and red cards are in the ratio 3:1. How many black cards do I have, and how many red cards?

(b) I have 25 coins in my pocket, either pennies or pounds. The ratio of pennies to pounds is 2:3. How many pennies and how many pounds are there?

7 Work out the following proportions.

Example: One company employs 12 men and 8 women. The other company employs 32 men and 18 women. Which company employs the greater proportion of women?

Solution: The first company employs 8 women out of 20 employees, which is 40%. The second company employs 18 women out of 50 employees, which is 36%. So the first company has a greater proportion of women.

(a) One class contains 25 students, 13 of which are from overseas. Another class has 40 students, 20 of which are from overseas. Which class has the highest proportion of overseas students?

(b) One internet forum has 28 guests and 12 registered users. Another forum has 21 guests and 9 registered users. Which forum has the highest proportion of guests?

5 Decimals, decimal places and significant figures

Using decimal numbers and rounding numbers to a certain number of decimal places

Fractions are one way of representing parts of a number – now we will look at decimals which is another way of representing parts of a number. Often in life we find ourselves in a situation where we want to 'approximate' a number to an appropriate degree of accuracy, which is where decimal places and significant figures come in.

Key topics

→ Decimals
→ Converting between fractions and decimals
→ Decimal places
→ Significant figures

Key terms
tenths hundredths decimal decimal places significant figures

An alternative way of representing numbers that aren't whole numbers is to use *decimals*. You have probably seen this when you use your calculator – your answer comes out in 'decimal' form. You also see this in your bank account – your balance might be £123.45, which is 123 pounds and 45 pence.

→ Simple decimals

Decimals are written as a whole number, followed by a decimal point (looks like a full stop) and then some more numbers. We'll do it first for very simple decimals and then move on.

When we write numbers, we talk about 'ones, tens, hundreds, thousands, ...'. In the decimal part (what comes after the decimal point), we talk about 'tenths, hundredths, thousandths, ...'. What does this mean?

Consider the fraction $\frac{3}{10}$. This is '3 bits out of 10' or 'three-tenths'. As a decimal you would write this as 0.3 – there is no 'whole bit' and there are 3 tenths.

Similarly, what is the fraction $5\frac{7}{10}$? This is 5 whole bits, and then 7 tenths, and so you would write this as 5.7 as a decimal.

Now think about the fraction $3\frac{1}{2}$. We have 3 whole bits, which is fine, but how many tenths do we have? Well, we know that $\frac{1}{2} = \frac{5}{10}$. So a half is the same as 5 tenths. Thus, we would write this fraction as 3.5 as a decimal.

Note that often we would write $3\frac{1}{2}$ as $\frac{7}{2}$ – so if you are given a fraction like this in top-heavy form, then you have to convert it to mixed form as we did in the fractions chapter.

→ More decimals

What about a number like $\frac{1}{4}$? You can't write this as a number of tenths, but you can write it out of 100, since $\frac{1}{4} = \frac{25}{100}$. So as a decimal, instead of just using tenths, we use hundredths as well: $\frac{1}{4}$ is the same as 0.25.

As another example, what is $\frac{15}{4}$ as a decimal? Well, this is the same as $3\frac{3}{4}$ and so we have 3 whole bits. And $\frac{3}{4} = \frac{75}{100}$. So this fraction is equivalent to 3.75 as a decimal.

To go further, a fraction like $\frac{317}{1000}$ is the same as 0.317 as a decimal, and so on.

What about $\frac{1}{3}$? No matter how hard you try, you won't be able to write this as tenths, hundredths, thousandths, and so on. You can get close – its *almost* the same as $\frac{33}{100}$ or $\frac{333}{1000}$ but it's not exactly the same. If you do it on your calculator, you will get the answer 0.33333333333 (the number of 3's depends on how many digits your calculator works to). This is only an approximation, the 3's carry on for ever.

You might have seen the recurring notation $0.\dot{3}$ to mean 0.3333333333 ...? The dot above the 3 means just keep writing it. However, this notation is generally discouraged – you might as well just write $\frac{1}{3}$!

Other decimals, you can't even write using recurring notation. You may have seen the number π which is 3.14159265358979323846 264338327950288419716939937510 ... and keeps going for ever, seemingly randomly. For most of our calculations, we don't need to be this precise, if you have had to use π in a formula you have probably used 3.14 or something similar. We will see what to do next.

smart tip

> Make a note every time you encounter decimals. This might be when using money to pay for things in the shops (e.g. a bill of £26.31), figures on the news such as inflation rates etc. (e.g. a rate of 3.2). By seeing them around you all the time, you realise their importance.

→ Decimal places

If our calculations don't have to be absolutely precise and accurate, but 'good enough' then we would tend not to use so many decimal numbers. For example, suppose the interest on your bank account was due to pay you £12.5123. This would obviously be considered as £12.51 for the purposes of your banking – the coins don't exist to make such a precise amount.

Similarly, when you take your temperature, your actual temperature might be 38.4983 ..., but you'd probably record this as 38.5 degrees. And, just as stated above, when you use π to work out a calculation, you might use 3.14 or some other approximation.

What you are doing is 'approximating' the number to something that's good enough for our purpose.

The number of digits after the decimal point is called the *number of decimal places.* The abbreviation *dp* is usually used for decimal places. So for example,

38.5 is to 1dp

3.14 is to 2dp

2.141 is to 3dp

5.0000 is to 4dp

The last one is particularly worth noting: this number is the same as 5, but we have written it as 5.0000 – there are 4 numbers after the decimal point, so 4 decimal places.

A number like 53 is to 0dp – there are no numbers after a decimal point (there isn't even a decimal point at all!).

When you are 'rounding' a number to some decimal places like this, remember that you are getting it as close to the number as you need. So, for example, to 1dp, what is 1.4999999? This is obviously closer to 1.5 than 1.4, so you would round this to 1.5.

There is a precise rule to follow, to round a number to a certain number of decimal places, which split into various different cases. The first case is easy!

Case 1a

If the number already has exactly the number of decimal places you need, then just leave it as it is! 4.78 to 2dp is just 4.78

Case 1b

If the number actually has fewer decimal places than you are asked for, then just add 0s to the end. So, 4.78 to 4dp is 4.7800

The examples that need more thought are when you are doing proper rounding, so the number of decimal places in the number is more than you actually want.

Case 2

In these cases, take the number, and simply write it down to as many decimal places as you need. Then look at the next digit in the original number. If it is less than 5, then you've finished. If it is 5 or more, then increase the last digit in your 'rounded' answer by 1. This is much clearer by example:

What is π to 2dp and 3dp?

Remember that π is 3.14159265358979323846264338327950288419716939937510 . . .

- To do this to 2dp:
  - Write down the number (π) with 2 decimal places which is 3.14
  - Look at the next digit in the number (π) – it's 1.

  - This is less than 5, so keep things just as they are
  - The answer is 3.14
- To do this to 3dp:
  - Write down the number (π) with 3 decimal places which is 3.141
  - Look at the next digit in the number (π) – it's 5.
  - This is 5 or more, so increase the last digit by 1
  - The answer is 3.142

This covers almost of all of the situations you will encounter apart from one.

Case 3

What is 3.497503 to 2dp? If we follow the rules from above then we write 3.49, look at the next digit which is 7, and increase 9 by 1. But this gives 10, and we can't write '10' – the answer 3.4 '10' doesn't make sense. The rule in this case is:

- Write it down to the appropriate number of digits as before
- Change the last 9 to 0
- Look at the *previous* digit and increase that by 1
- Keep going if you need to

So, 3.497503 to 2dp? Follow the rule.

- Write down the number with the first 2 decimal places – you get 3.49
- Look at the next digit, it's 7
- So you need to increase the 9 by 1. This needs to follow the Case 3 rules.
- Change the 9 to 0
- Look at the previous digit, it's 4. So increase this by 1 to get 5.
- You get 3.50

The point is, that 3.50 is the closest number you can get to 3.497503 with two decimal places – think about it.

Remember the basic rule:

- **Write down the number to as many decimal places as you need.**
- **If the next digit is a 'small number' 0 to 4 then leave things as they are.**

- If it is a 'large number' 5 to 9 then you need to increase the last digit you wrote down by 1.
- You need to pay special attention if that digit is a 9 by making it 0 and increasing the previous digit by 1 instead.

→ Significant figures

In the last section, we talked about approximating decimal numbers to a 'reasonable' accuracy for our purposes. We often do the same sort of approximating of big numbers, too, to give them to 'enough accuracy that we need'. For example, if an atomic particle is moving at 320021 m/s, you'd probably say it was going at about 320000 m/s. If you won £11102.53 in the lottery and someone asked how much you won, you'd probably say you won 'about £11000'. The number of digits we write before rounding off is called the number of *significant figures* (sf).

Significant figures with whole numbers

First of all, look at approximating a whole number (positive integer) to a certain number of significant figures.

Case 1

If the number of digits in the whole number is the same as the number of significant figures required, then just write the number down. So, for example, 536 to 3sf is just 536.

Case 2

If the number of digits is more than the number of significant figures required then we act somewhat similarly to decimal places. Take the number, and write down the number of significant figures you need. Then look at the next digit. If it is less than 5, then leave the rounded number as it is. If it is 5 or more, increase the last digit of the rounded number by 1. *In either case, finally, fill in 0s so the rounded number is the same length as the original number.* Again let's clarify this with an example.

We are going to write the number 142981 to 2sf and 3sf.

- To do this to 2sf:
  - Write down the first 2 digits, giving 14

  - Look at the next digit in the number – it's 2
  - This is less than 5, so leave the rounded number as it is
  - Now fill in 0s to make it the right length – it should have 6 digits, so we write 140000
- To do this to 3sf:
  - Write down the first 3 digits, giving 142
  - Look at the next digit in the number – it's 9
  - This is 5 or more, so increase the last digit of our rounded answer by 1, giving 143
  - Now fill in with 0s to make it the right length – it should have 6 digits, so we write 143000

Why do we have to write down the 0s? The point is we are approximating the number. If you had £17026 in your bank account and I rounded that to 'about £17' you wouldn't be very happy – the correct rounding is 'about £17000'.

Case 3

The last case in this section is, just like with decimals, the awkward situation when the digit we want to increase by 1 is 9. We follow exactly the same procedure as with decimals – we change it to 0 and increase the previous digit by 1 instead.

So, what is 398232 to 2sf?

- Write down the first 2 digits – 39
- Look at the next digit, it's 8
- So we need to increase the 9 (last digit of the 'rounded' number) by 1.
- Remember, to do this, change it to 0, and then increase the previous number by 1, giving us 40.
- Now fill in the 0s to make it the right length – the answer is 400000

This seems slightly odd – this only appears to be to 1sf? But the point is that the number *is* closer to 400000 than 390000 or 410000, say, so this is the correct rounding.

Don't forget – all we are doing is rounding again, to the nearest number that is accurate enough for our purposes.

Significant figures with decimals

Significant figures can be used equally well when decimals are involved.

Case 1

First, if you are given a decimal number and asked to round it to more significant figures than are in the actual number, then add an appropriate number of 0s to the end. So, for example, 3.1 to 4sf is 3.100 (four digits given).

Case 2

In almost all other cases you follow the rule of writing down as many digits as you need and then rounding in our usual way.

For example, what is 5.2497 to 2sf, 3sf and 4sf?

- To 2sf:
  - write down the first 2 digits, giving 5.2
  - Look at the next digit, it's 4 so leave things as they are
  - That's the answer to 2sf: 5.2 (2 significant figures)
- To 3sf:
  - Write down the first 3 digits, giving 5.24
  - Look at the next digit, it's 9, so we need to increase the last digit of our rounded answer by 1, so change the 4 to 5
  - We get the answer 5.25 (3 significant figures)
- To 4sf:
  - Write down the first 4 digits, giving 5.249
  - Look at the next digit, it's 7, so we need to increase the last digit of our rounded answer by 1, unfortunately this is a 9
  - Our rule says change it to 0 and go back to the previous digit and increase that 1 instead. We get 5.250 (4 significant figures)

Case 3

There is just one more case to consider. When our number is very small (almost zero) and of the form 0.0000 ... then we only start counting the significant figures when we reach the first non-zero digit.

So, for example, what is 0.000024378 to 2sf and 3sf? We don't start counting the significant figures until we reach the first non-zero digit which is 2. So to 2sf, this is 0.000024. To 3sf it would be 0.0000244, using our usual rounding rules.

Remember – significant figures are all about rounding the number to a degree of accuracy that we need.

Try not to get decimal places and significant figures confused – although they are related concepts, they are different things.

Decimal places always start 'counting' from the decimal point – significant figures start counting from the first 'important' number.

So, as an example, £2341.4619 is £2341.46 to 2dp but £2300 to 2sf. And 0.0000002321 is 0.00 to 2dp but 0.00000023 to 2sf.

smart tip

Lots of times that you encounter numbers, they are rounded like this. For example, if you are driving, your actual speed might be 30.0268749 ... mph but you'd obviously just say you were driving at 30mph. Every time you come across numbers, think as to how they have been rounded, and how annoying it would be if we didn't round off (imagine street signs with a speed limit of 30.0268749 mph)!

Summary

Often there's a need to approximate numbers. You can't make £23.6247 pence, so you need to approximate it to £23.62. But you need to approximate it fairly – £23.769 is closer to £23.77 than it is to £23.76. You also often need to approximate figures to summarise them – you often hear on the news about million-pound deals (for a footballer say) – the actual figures involved are quite complex, say £25001564, which in reality would be reported as £25000000 (25 million pounds) just to make things simpler.

Exercises

1 Convert the following fractions into decimals. Remember that for top-heavy fractions you will need to convert them into mixed fractions first.

Example: $\frac{1}{5}$

Solution: $\frac{1}{5} = \frac{2}{10}$ which is 0.2 is a decimal.

(a) $\frac{7}{10}$ (b) $\frac{3}{5}$ (c) $\frac{1}{4}$ (d) $\frac{3}{50}$

(e) $\frac{11}{500}$ (f) $\frac{11}{2}$ (g) $\frac{23}{5}$ (h) $\frac{17}{4}$

2 What are the following decimals, as fractions in their lowest terms?

Example: 0.6

Solution: $0.6 = \frac{6}{10} = \frac{3}{5}$ cancelling to its lowest terms.

(a) 0.3 (b) 0.8 (c) 0.25 (d) 0.48

(e) 2.3 (f) 4.5 (g) 0.33 (h) 0.01

3 Write the following numbers to 3 decimal places (3dp).

Example: 5.31283

Solution: Write down the decimal places 5.312. Look at the next digit, which is 8. This is one of our large numbers (5 or more) so we need to increase the last digit of our answer by one, to get 5.313

(a) 5.386189 (b) 4.62572 (c) 0.07918 (d) 0.19652

(e) 2.892989 (f) 3.07982 (g) 2.28972 (h) 0.8

4 Give the following numbers to 3sf (3 significant figures):

Example: 526431

Solution: Write down the first 3 digits, to get 526. The next digit is 4, which is a 'small' number (less than 5), so leave things as they are. Now fill in the appropriate number of 0s to get 526000.

(a) 23125 (b) 1289 (c) 219532 (d) 399999

(e) 1.2537 (f) 12.871 (g) 0.379526 (h) 0.00013861

6 Scientific notation

Representing very small and very large numbers using scientific notation

Key topics

→ Scientific notation
→ Multiplication and division with scientific notation

Key terms
scientific notation coefficient exponent

→ Scientific notation

Now let's think about writing very big or very small numbers. In an example we saw in the previous chapter, it's a bit unwieldy to write 0.000000231 with all the zeros. The speed of light is about 300000000m/s. The trouble with writing all these zeros down is that it's very easy to make mistakes, you have to be sure you counted the right number. And suppose someone looks at a complex formula with all these 0s everywhere, the chance of an error somewhere is huge.

To take things to ridiculous proportions, there is a number called a googol which you may or may not have heard of (it became quite famous due to the quiz show Who Wants To Be A Millionaire, when it was the final million-pound question in the 'coughing' scandal, if you remember that). Here it is:

100000000000000000000000000000000000000000000000000
00000000000000000000000000000000000000000000000000

that's 1 with a hundred 0s after it. It's not that useful a number, although it is used to emphasise really big things (it's bigger than the number of atoms in the known universe, which is quite a lot when you think about it).

Now, imagine if you used this in a calculation, what are the chances of you not quite counting it right when you count the 0s? Quite high, I'd imagine. But if you do make a mistake, your calculation is all wrong.

The idea is to express our numbers in an easier format which makes the number obvious. We'll look at very big numbers first, and then very small numbers.

smart tip

Examples like the above, where maths things you might never have thought you needed to know suddenly appear, can make the subject more interesting and real. If you are on Who Wants To Be A Millionaire one day and there's a maths question, wouldn't it be great if you knew the answer? So notice where maths appears in such situations and note it!

→ Very big numbers

Look at the following pattern:

$$5 = 5$$

$$50 = 5 \times 10$$

$$500 = 5 \times 10 \times 10$$

$$5000 = 5 \times 10 \times 10 \times 10$$

$$50000 = 5 \times 10 \times 10 \times 10 \times 10$$

$$500000 = 5 \times 10 \times 10 \times 10 \times 10 \times 10$$

and so on. In each case you can write the number as 5, multiplied by as many 10s as there are 0s after the number. The notation for something like $10 \times 10 \times 10 \times 10$ where there are many of the same thing multiplied together is to write it as 10^4 (we'll see this later when we deal with indices). This means take the number 10, and multiply four of them together.

So you could write 50000 as 5×10^4. Similarly you could write, say, 300000 as 3×10^5.

Our 'googol' above was 1 with a hundred 0s after it. So this is the same as 1×10^{100}.

What about something like 320000? You could write this as 32×10^4.

But this isn't the conventional way to write it. The conventional way is to make sure the *coefficient* – which we will refer to informally as the 'main' number at the front – is always a number between 1 and 10 (not including 10).

Note that $32 = 3.2 \times 10$.

So $320000 = 32 \times 10^4 = (3.2 \times 10) \times 10^4 = (3.2 \times 10) \times 10 \times 10 \times 10 \times 10 = 3.2 \times 10^5$

and this is in the form we require (the coefficient 3.2 is between 1 and 10).

This is called writing a number in *scientific notation*. I'll give you a basic rule to follow for big numbers, using 5340000 as an example:

- Write down the big number, in this case 5340000.
- If it is a whole number, imagine it to have a decimal point at the end: 5340000.
- Count how many places you would have to 'move' the decimal point along to the left, to leave a number between 1 and 10 (but not 10).
  - In this case, if you move the decimal point 6 places to the left you get 5.340000.
- Ignoring any 0s left at the end after the decimal point (they do not serve any useful purpose), write down this number obtained, then multiply by 10^d where d is the number of places you moved the decimal point by.
  - In this case, you moved it 6 places so you write down 5.34×10^6.

On your calculator, you can check that these two are exactly the same thing.

The number d (the number above the 10) is known as the *exponent*.

Note that if it is has a decimal part, say we have 5340000.576, you still move the decimal point in the same way (6 places to the left) and you end up with 5.340000576×10^6.

→ Very small numbers

Sometimes we deal with very small numbers close to 0. For example when dealing with atomic particles, we might have numbers like

0.00000006823 when measuring mass and so on. Again, we need to be careful with counting the number of zeros correctly.

We follow a similar procedure to that we did with big numbers, but in some sense 'in reverse'.

- Write down the small number, in this case 0.00000006823.
- Count how many places you would have to 'move' the decimal point along to the *right*, to leave a number between 1 and 10 (but not 10)
  - In this case, if you move the decimal point 8 places to the right you get 6.823
- Write down this number obtained, then multiply by 10^{-d} (*note the minus sign)* where *d* is the number of places you moved the decimal point by.

In this case, you moved it 8 places so you write down 6.823×10^{-8}.

Remember – for big numbers, move the decimal point to the left and the power of 10 will be positive. For small numbers, move the decimal point to the right and the power of 10 will be negative.

→ Multiplying numbers in scientific notation

If you want to multiply two numbers written in scientific notation, there is an easy way. Say you want to multiply 2×10^4 by 3×10^5. Note that this is the same as

$$(2 \times 10 \times 10 \times 10 \times 10) \times (3 \times 10 \times 10 \times 10 \times 10 \times 10)$$
$$= (2 \times 3) \times (10 \times 10 \times 10 \times 10 \times 10 \times 10 \times 10 \times 10 \times 10)$$

since it doesn't matter which order we do multiplication in. And this is the same as 6×10^9. The general rule to follow is:

- Multiply the two 'main' numbers (coefficients) together – in this case we multiplied 2 and 3 together to get 6.
- *Add together* the powers of 10 (exponents) – in this case we added together 4 and 5 to get 9.
- Make sure the answer you get is in scientific notation.

As another example, what is $(3 \times 10^6) \times (4 \times 10^2)$?

We multiply the coefficients (3×4) and add the exponents together $(6 + 2)$ to get 12×10^8. But this isn't in scientific notation, because

12 is too big – we need a number between 1 and 10. So, move the decimal point one place to the left, which increases the power of 10 by 1, and we have the answer 1.2×10^9.

It works with negative powers as well. For example, what is $(3 \times 10^9) \times (2 \times 10^{-5})$?

Following exactly the same rules, we multiply 3×2 to get 6, and add the powers of 10: this gives $9 + (-5)$ which is 4. So we get 6×10^4.

→ Dividing numbers in scientific notation

To divide two numbers written in scientific notation, we follow a similar procedure but instead we:

- *Divide* the main numbers.
- *Subtract* the powers of 10.

So, for example, what is (3×10^8) divided by (2×10^5) in scientific notation? (You would usually write this as a fraction $\frac{3 \times 10^8}{2 \times 10^5}$)

Dividing 3 by 2 gives 1.5 as a decimal, and subtracting the exponents gives $8 - 5 = 3$. So the answer is 1.5×10^3.

Again, make sure you do give answers in scientific notation:

What is $\frac{3 \times 10^9}{6 \times 10^4}$ in scientific notation?

Dividing 3 by 6 we get 0.5, and subtracting the exponents we get $9 - 4 = 5$ so this is 0.5×10^5. But this isn't scientific notation because 0.5 is too small, we need a number between 1 and 10. We need to move the decimal point one place to the right, which means decreasing the power of 10 by 1. So we get 5×10^4.

And, again, it works with negative numbers too. What is $\frac{9 \times 10^7}{2 \times 10^{-5}}$?

Dividing 9 by 2 gives 4.5, and subtracting the powers we get $7 - (-5)$ which is 12 (be careful with minus signs!). So this is 4.5×10^{12}.

It is always worth checking your answer to see if it looks right – in this case we are dividing a big number by a small number, so we'd expect an even bigger number (there are lots and lots of small things dividing into a big thing), which is what we got.

Get used to scientific notation by converting every very big (or very small) number you see into scientific notation. Next time you read of a footballer being transferred for £40 million, or Government spending being £300 million, say, think about what these numbers mean and write them in scientific notation. Write them out in full as well (£300 million is £300000000) and realise that it can get quite confusing with all the zeros, and scientific notation can make it a lot easier!

Summary

Scientific notation is a useful way to write down very small or very large numbers. Instead of worrying about all the 0s and wondering if we wrote the right number down, then the exponent (the number above the 10) tells us how big or small the number is. This is very useful in large calculations – rather than write down the really big numbers and spend ages dealing with them, we just have to deal with the small coefficients and exponents, which are much easier to deal with and save us lots of time – in today's fast-paced, hi-tech world, time is precious!

Exercises

1 Write the following numbers in scientific notation.

Example: 340000

Solution: Imagine it to have a decimal point at the end. You would need to move the decimal point five places to the left to get 3.4, so the answer is 3.4×10^5.

(a) 20000	(b) 23400000	(c) 5600000	(d) 130000000000
(e) 153	(f) 10000000	(g) 1932.486	(h) 7

2 Write the following numbers in scientific notation.

Example: 0.00078

Solution: You need to move the decimal point four places to the right to get 7.8, so the answer is 7.8×10^{-4}.

(a) 0.0036	(b) 0.000079	(c) 0.00000999	(d) 0.00178
(e) 0.0000000078	(f) 0.00000001	(g) 0.1	(h) 0.134

3 Work out the following multiplications, leaving your answers in scientific notation.

Example: $(3 \times 10^5) \times (3 \times 10^6)$

Solution: Multiply together the coefficients (main numbers) to get $3 \times 3 = 9$, and add together the exponents (powers of 10) to get $5 + 6 = 11$, so you have 9×10^{11}.

(a) $(4 \times 10^6) \times (2 \times 10^4)$

(b) $(2 \times 10^3) \times (2 \times 10^{15})$

(c) $(2 \times 10^{12}) \times (3 \times 10^{-8})$

(d) $(2 \times 10^6) \times (4 \times 10^{-12})$

(e) $(5 \times 10^8) \times (7 \times 10^7)$

(f) $(6 \times 10^{-4}) \times (4 \times 10^{-5})$

4 Work out the following divisions, leaving your answers in scientific notation:

Example: $\frac{9 \times 10^7}{3 \times 10^2}$

Solution: Divide the coefficients (main numbers) to get $\frac{9}{3} = 3$, and subtract the exponents (powers of 10) to get $7 - 2 = 5$, so you have 3×10^5.

(a) $\frac{8 \times 10^{11}}{4 \times 10^3}$

(b) $\frac{9 \times 10^9}{2 \times 10^4}$

(c) $\frac{6 \times 10^7}{2 \times 10^{-5}}$

(d) $\frac{5 \times 10^{-8}}{4 \times 10^3}$

(e) $\frac{2 \times 10^{10}}{8 \times 10^3}$

(f) $\frac{3 \times 10^{-7}}{3 \times 10^{-8}}$

5 Calculate the time taken in the following questions using scientific notation.

Example: A computer program takes 0.0003 seconds to run. How long does it take to run the program 2000000 times?

Solution: 0.0003 is 3×10^{-4} and 2000000 is 2×10^6. So the time taken is $(3 \times 10^{-4}) \times (2 \times 10^6) = 6 \times 10^2$ using the usual rules. This is the same as 600 seconds.

(a) An atomic particle takes 0.00004 seconds to travel 1 metre. How long will it take to travel 2000000 metres?

(b) A specialised program takes 0.02 seconds to find a particular piece of data. Giving your answer in scientific notation, how long will it take to find 65000000 pieces of data?

Note: 65000000 (or 65 million) was chosen for (b) as it is a reasonable approximation for the population of the UK.

7 Indices

Writing repeated multiplications easily, roots and the laws of powers

In Chapter 6, when we looked at scientific notation, we saw the convention to use for example 10^4 to stand for $10 \times 10 \times 10 \times 10$, so 10^4 is shorthand notation for 4 tens all multiplied together. This concept generalises – here we will investigate this notation and introduce some laws that you should know that can make our calculations simpler.

This topic is formally referred to as *indices* (the plural of *index*), but most people refer to it as 'powers' or even 'orders' (this is the missing 'O' in BODMAS!).

Key topics

→ Powers
→ Roots
→ Power laws

Key terms
power base index square square root power laws

→ Basic indices

Something like 10^4 means four tens all multiplied together, so $10 \times 10 \times 10 \times 10$ (which equals 10000 if you work it out). This can be read as '10 to the power 4'

In an expression like this, the 'big number' 10 is called the *base* and the small number, written up in the air above the 10, is called the *index* or *power*.

Similarly, with 2^5, the base is 2 and the index is 5, and can be read as 2 to the power 5. This is the same as $2 \times 2 \times 2 \times 2 \times 2$ (five 2s multiplied together), which you can check is 32.

It is quite common to have an index of 2, which is called a *square*. For example, 6^2 means 6×6 (two 6s multiplied together) which is 36.

As square numbers are quite common in maths, it is worth knowing the basic ones

$$1^2 = 1 \times 1 = 1$$
$$2^2 = 2 \times 2 = 4$$
$$3^2 = 3 \times 3 = 9$$
$$4^2 = 4 \times 4 = 16$$
$$5^2 = 5 \times 5 = 25$$
$$6^2 = 6 \times 6 = 36$$
$$7^2 = 7 \times 7 = 49$$
$$8^2 = 8 \times 8 = 64$$
$$9^2 = 9 \times 9 = 81$$
$$10^2 = 10 \times 10 = 100$$
$$11^2 = 11 \times 11 = 121$$
$$12^2 = 12 \times 12 = 144$$

When the power is 3, such as in 2^3 (which is $2 \times 2 \times 2 = 8$) then this is referred to as a *cube*. These are less common so you don't have to try to learn these!

smart tip

It is worth the effort to learn some things like this. As with times tables, you might encounter them quite often and it's really a boost to be able to do them quickly in your head without needing a pen and paper or a calculator, and it can save you time in exams!

→ Other indices

Defining something like 2^4 as 'four 2s multiplied together' is fine and gives us a good useful definition when the index is some nice positive number. It extends fairly easily to when the index is 1; 2^1 is just 'one 2' which is just 2. So anything to the power 1 is just that number, for example $4^1 = 4$ and $63^1 = 63$.

It gets a bit harder when the index isn't a normal nice positive whole number. Something like 2^0 or 2^{-3} isn't easy to explain: how can you multiply -3 lots of 2 together? It just doesn't make sense conceptually.

Still, it would be nice to define it in some way, so what we are going to do is make a definition that makes sense and allows us to use basic mathematical laws without any problems. Essentially, here, we are defining something the way it is because it makes things easier!

Look at the following table that gives all the powers of 2 from 1 to 6.

2^1	2^2	2^3	2^4	2^5	2^6
2	4	8	16	32	64

Notice that each answer is twice what it was before for the previous power – alternatively looking backwards, that every answer is half of the next one.

What should 2^0 be? If you were to continue this table to the left, 2^0 would be half of the next value, which is half of 2, which is 1.

So it would make natural sense to define 2^0 to be 1, so that our table would continue in a nice pattern. Remember, we are defining this because it fits with what's already clear – we're trying to make things easy!

The same example works for any number – so we are going to say that for any number, if the index is 0, then the answer is 1. This will fit nicely into the system we are developing. You can learn this as

Anything to the power 0 is 1.

So for example, $20^0 = 1$, and $3548245625^0 = 1$, and so on. The only exception is 0^0 which we can't define – like we can't define dividing by 0 – but don't worry about this, just learn the rule that anything to the power 0 gives the answer 1.

→ Negative powers

Having dealt with 0, the next question is what to do with negative powers. What on earth is 2^{-3} meant to mean?

Suppose we took our table from before (powers of 2) and tried to follow the pattern that every power is half of the previous one. Then we ought to have a table like this:

2^{-4}	2^{-3}	2^{-2}	2^{-1}	2^{0}	2^{1}	2^{2}	2^{3}	2^{4}	2^{5}	2^{6}
$\frac{1}{16}$	$\frac{1}{8}$	$\frac{1}{4}$	$\frac{1}{2}$	1	2	4	8	16	32	64

So, it would be nice to adopt a rule for negative powers that follows this rule. Note that, for example, 2^{-3} is $\frac{1}{8}$, which is the same as $\frac{1}{2^3}$. This leads us to a general rule:

Anything to a negative power is simply 1 divided by the number to the positive power.

So for example, 5^{-8} is the same as $\frac{1}{5^8}$, 3^{-10} is the same as $\frac{1}{3^{10}}$ and so on. Most of the time, you won't be able to actually calculate these, so you can leave your answers like that, but if you are given something like 2^{-3}, this is the same as $\frac{1}{2^3} = \frac{1}{8}$.

→ Power laws

What we are going to do now is look at some different rules that follow from the basic definitions we made so far. These will form what are called the 'power laws'. We'll do these fairly informally for now, and formalise later on in the logarithms chapter after we have discussed algebra.

Multiplying

Suppose you have something like $2^3 \times 2^4$. What is this?

Well, 2^3 means $2 \times 2 \times 2$, and 2^4 means $2 \times 2 \times 2 \times 2$. So in total we have $(2 \times 2 \times 2) \times (2 \times 2 \times 2 \times 2)$ which is seven 2s multiplied together, so 2^7.

Note the 7 comes from adding together the 3 and the 4 (three 2s and then another four 2s). So $2^3 \times 2^4 = 2^7$, where the 7 has come from adding up the 3 and the 4.

This always works as long as the base is the same. So for example

$4^5 \times 4^6 = 4^{11}$ since $5 + 6 = 11$

And it also works for negative numbers (just add the indices) so, for example,

$3^8 \times 3^{-3} = 3^5$ since $8 + (-3) = 5$

It only works when the base is the same though! You can't do anything with something like $2^3 \times 3^4$.

The basic rule to remember is:

As long as the base is the same, then in a multiplication you can just add the indices together.

Dividing

What is $\frac{2^7}{2^3}$? Writing this in full, this is $\frac{2 \times 2 \times 2 \times 2 \times 2 \times 2 \times 2}{2 \times 2 \times 2}$ and the 2s cancel to leave $2 \times 2 \times 2 \times 2 = 2^4$. Note that we 'cancelled' the three 2s on the bottom from the seven 2s on the top – basically to get the final index of 4 we subtracted the bottom power from the top.

This works in general as well, so for example $\frac{4^{12}}{4^5} = 4^7$, since the 7 is obtained by subtracting the powers $12 - 5$.

Again, it works for negative powers as well, so for example $\frac{3^5}{3^{-6}} = 3^{11}$ since $5 - (-6) = 11$.

Again it only works if the base is the same. The basic rule is:

As long as the base is the same, then in a division you can just subtract the indices (subtract the bottom from the top).

Power of a power

What is $(2^3)^4$? By definition, this is $(2 \times 2 \times 2)^4$, which means multiply four lots of $(2 \times 2 \times 2)$ together, giving $(2 \times 2 \times 2) \times (2 \times 2 \times 2) \times (2 \times 2 \times 2) \times (2 \times 2 \times 2)$, which if you count, is the same as 2^{12}.

Note that this is 'the power three, done four times', or in this example, the final power is $3 \times 4 = 12$. Hence, the general rule is:

If you have a power, to another power, then multiply the two indices together.

So for example, $(4^5)^3 = 4^{15}$ multiplying the two indices together.

Fractional powers

Finally, what should be the answer if you use an index that is a fraction, such as $4^{\frac{1}{2}}$?

Note from what we did before, that $4^{\frac{1}{2}} \times 4^{\frac{1}{2}}$ should be the same as 4^1 (adding the indices together) which is just 4. So $4^{\frac{1}{2}}$ should be defined as whatever number that when you multiply it by itself, gives 4. A little thought and you'll realise the answer should be 2, since $2 \times 2 = 4$.

Hence $4^{\frac{1}{2}} = 2$.

This is what is known as a *square root*. The square root of a number is a number that when multiplied by itself, makes the number you are asking about. This is the 'opposite' of squaring that we talked about before.

For example, $16^{\frac{1}{2}} = 4$ since 4 is the 'square root' of 16 (as $4 \times 4 = 16$).

The mathematical notation for square root is the symbol $\sqrt{}$. So we can write, for example, $\sqrt{16} = 4$.

Anything to the power $\frac{1}{2}$ is the square root of the number.

Further fractions get more complicated. You don't need to worry too much about these for now: something like $27^{\frac{1}{3}}$ is what is called a 'cube root' or 'third root' of 27 – a number that when you multiply it by itself, and then by itself again, gives 27, so 3 in this case. Don't worry about this for now – just know it exists!

You can of course combine all of these power laws together, so for example $\frac{(2^{\frac{1}{2}})^8}{2^{-3}} = \frac{2^4}{2^{-3}} = 2^7$

smart tip

Unfortunately for a topic like this, you really just have to learn the rules. But if you understand why we make the rules we do, it makes it easier to learn them! Always, in your studies, try to understand why things are defined the way they are – it's so much easier to learn something you understand!

Summary

This topic is a good example in maths of trying to make things easier. It makes perfect sense to define something like 3^{12} rather than go to all the effort of writing

$$3 \times 3 \times 3 \times 3 \times 3 \times 3 \times 3 \times 3 \times 3 \times 3 \times 3 \times 3$$

Also realise that while powers like these are natural, defining something like 3^{-5} isn't very natural – so we chose a definition for it that fitted into everything else we were doing. We defined zero, negative and fractional powers, so that they fitted nicely into the setup we already had.

We're just trying to make life easier for ourselves by defining things in the way we do. Remember – maths tries to make your life easier!

Exercises

1 Calculate the following powers – try to do them in your head, rather than use a calculator.

Example: 2^4

Solution: $2^4 = 2 \times 2 \times 2 \times 2 = 16$.

(a) 2^3 (b) 10^2 (c) 7^2 (d) 3^3

(e) 2^6 (f) 12^0 (g) 386745^0 (h) 2^{-2}

(i) 3^{-4} (j) $4^{\frac{1}{2}}$ (k) $64^{\frac{1}{2}}$ (l) $8^{\frac{1}{3}}$

2 Write the following as a single power.

Example: $2^4 \times 2^7$

Solution: Add the powers together to get 2^{11}.

(a) $2^3 \times 2^6$ (b) $8^7 \times 8^{15}$ (c) $2^8 \times 2^{-3}$ (d) $4^{-3} \times 4^{-7}$

(e) $\frac{5^8}{5^2}$ (f) $\frac{4^3}{4^{-2}}$ (g) $\frac{6^{-5}}{6^3}$ (h) $\frac{2^{-4}}{2^{-9}}$

(i) $(2^4)^3$ (j) $(3^2)^{-3}$ (k) $(4^8)^{\frac{1}{2}}$ (l) $(3^2)^0$

3 *Extra question – is it possible ever to get a negative number as an answer to an indices question like the ones above?*

ALGEBRA

8 An introduction to algebra

Using symbols to represent numbers and an introduction to algebra

'Algebra' is a word that puts fear into the hearts of a lot of people and conjures up terrifying thoughts of strict maths teachers talking in what seems like a different language. But you don't need to be scared of the word – all algebra is, basically, is abbreviating things which you do all the time anyway in your normal daily life.

Key topics

- → Using symbols
- → Creation and evaluation of formulae
- → Simplifying expressions

Key terms

algebra symbols evaluation collecting like terms simplification

→ What is algebra?

A lot of things you have done in mathematics so far has involved numbers. But many problems in all fields and your daily life are general problems which can be formulated in general terms.

For example, your employer gives all employees a Christmas bonus of £100. What is your December salary? If you normally earn £1000 a month then your salary will be £1100. If you normally earn £500 a month then your December salary will be £600. If you normally earn £2000 a month then your December salary will be £2100.

You could write this as a sort of equation:

'December salary' equals 'monthly salary' plus 'a hundred'

but to use long phrases like this isn't really very convenient. Algebra is all about using symbols to 'stand for' something.

Suppose we use the letter d to stand for 'December salary' and the letter m to stand for monthly salary. Then we could write:

$$d = m + 100$$

If you are told what m (your monthly salary) is, you can work out d (your December salary).

So, if $m = 1500$, what is d? Put $m = 1500$ in the formula:

$$d = 1500 + 100 = 1600$$

and so your December salary d is 1600.

The point is, you can use this formula for all the employees – they might all have different values of m but you can use this formula to work out their d.

Remember – the letters are just symbols to stand for something.

smart tip

Algebra is basically just using a letter to abbreviate something. You do this all the time, probably every day. You might say you are driving at 30mph, where mph is an abbreviation for 'miles per hour'. Similarly, text-message speak uses 'u' instead of 'you', and so on. Algebra is just using abbreviations to write things concisely – so think about how many times you use abbreviations every day and when you realise that is all algebra really is, it doesn't seem very scary!

→ Creation and evaluation of formulae

Often we are given a problem in words, like that above with the whole statement about December salary bonus payments. From this statement in words, we create a formula using symbols. The process of working out a specific value of the formula for particular values of the symbols is called *evaluation*. Let's look at an example:

Example

At a sale, all books are £5 and all posters are £3. Write down a formula that works out the total cost for a customer who buys some books and some posters. If I buy 3 books and 4 posters, what does it cost me?

Let b = the number of books bought (an abbreviation)
Let p = the number of posters bought (an abbreviation)

Then, the books cost £5 each, so the books in total cost the customer $5 \times b$.

> In algebra, we don't usually write down the multiplication sign in expressions like this (the reason being that we use x a lot in algebra, and an x looks like a multiplication sign, so we can get confused). We would abbreviate this to $5b$ – when you see an expression like this, be aware it is short for $5 \times b$. Similarly, something like $2xy$ is short for $2 \times x \times y$ and so on.

Similarly, the posters cost £3 each, so the posters in total cost $3 \times p = 3p$. So to work out a customer's total bill we add these two numbers up:

$$5b + 3p$$

This is the formula. In my example, I bought three books and four posters. So for me, $b = 3$ and $p = 4$.

So, evaluating, the cost is $5 \times 3 + 3 \times 4 = 27$ (remember the BODMAS rule!)

My purchases cost me £27.

What's the point of doing this? Well, if cashiers just build the formula into their till machine, then they can work out the bill for every customer who comes along with a different number of posters and books, just using the formula $5b + 3p$, rather than manually working it out every time. This is clearly going to save time and effort. Of course, real-life formulae are usually more complicated, but hopefully you get the idea.

The formulae involved can get quite complicated, but remember that to evaluate them, it's simply a case of putting the appropriate values in. Our formulae won't get too complicated here, but even for complex formulae like in rocket science, to evaluate them, it is still just a case of putting values in.

Example

Evaluate $\frac{5a + 3}{b - 1}$ when $a = 5$ and $b = 8$.

Just put the numbers in: if $a = 5$ and $b = 8$ then this is $\frac{5 \times 5 + 3}{8 - 1}$ which is $\frac{25 + 3}{7} = \frac{28}{7} = 4$

and so the answer is 4.

→ Simplifying expressions

Sometimes complicated expressions can be made simpler.

Suppose I have 2 essays to write, and then I suddenly have another 9 essays to write. It's clear that I have 11 essays to write in total. In algebraic terms:

Suppose I have the expression $2x + 9x$, where x stands for some quantity (in this case essays). Here I've got 2 of x, and then another 9 of x, so clearly I've got $11x$ in total.

So $2x + 9x$ is the same as $11x$.

Don't forget that it doesn't matter which order you do addition in.

For example, take the expression $3x + 4y + 2x + 3y$. We have $3x$ and another $2x$, so $5x$ in total. Similarly we have $4y$ and another $3y$, so $7y$ in total. Hence this is the same as $5x + 7y$.

Note that it doesn't matter what order you write things in – if you were to write $7y + 5x$ for example, that's perfectly OK.

Note that different symbols represent different things. If you had 3 essays, then were given 4 tests, then another 2 essays, then another 3 tests, then you've got 5 essays and 7 tests to do – you would probably say it like that ('Agh, I've got 5 essays and 7 tests to do!' rather than 'Agh, I've got 12 things to do!') – the essays and tests are distinguished.

As another example, take $7x - 2y - 3x + 5y$. What is this the same as? Well, we have $7x$ and then later on $-3x$. So in total there are $4x$ (we start with 7 xs, and take 3 of them away). Similarly we have $-2y$ and then later $5y$ so in total we have $3y$. Remember the rules for negative numbers (if you have forgotten these revise Chapter 2).

This is an example of what is known as *collecting like terms.* (Here, a *term* is one of the things being added up in an algebraic expression, so 'like terms' refers to things that are 'alike', so what you are doing is collecting together things that are alike, they have the same letters in them). It's always good to have your expressions as simple as possible – called *simplifying* them.

Always look for ways to simplify your expressions. For example, look at the complicated-looking expression:

$$\frac{3x}{2} - \frac{3x}{4} + \frac{x}{2}$$

Note the convention in the last term – there is no number before the symbol x, this is a shorthand way of writing $1x$, in the same way we might just say I have to do 'an essay' rather than 'one essay'.

As in the fractions chapter, make all the fractions have the common denominator 4: this is the same as

$$\frac{6x}{4} - \frac{3x}{4} + \frac{2x}{4}$$

which is the same as

$$\frac{6x - 3x + 2x}{4} = \frac{5x}{4}$$

which is much simpler, I'm sure you'd agree.

smart tip

Simplifying things makes them easier. As humans, we don't like complicated things, we just want to make things nice and simple. All we are doing with algebra here is taking something complicated and trying to make it easier. You do this all the time in your life – think about trying to make the instructions you give as simple as possible – this is all we are doing when simplifying: why write something complicated when we can write something easier? Think about your daily life and how you try to make things easier in everything you do!

→ Simplification of more complex terms

When you collect things together like above, the basic rule is that you can *only* collect together things that have exactly the same letters in them.

For example, if you have $2xy + 3xy$ then this is '2 lots of xy' and another '3 lots of xy' which is in total $5xy$.

However, suppose you had the expression $2xy + 3xz$. There is nothing you can do to make this simpler (at least for now, though you will see something called factorisation later which can help). The letters in the terms are different, yes both terms contain an x, but xy is different from xz.

So always remember that you must only ever collect together things where the letters are *exactly* the same.

For example, what is $2xy + y + 3x + 7 - 7xy + 5x - 2$?

There is a $2xy$ and later a $-7xy$ which in total is $-5xy$. There is a $3x$ and later a $5x$ so that's $8x$ in total. There is a positive 7 and a -2, so we get a 5. Finally there is a single y so it just stays as it is. So you get $-5xy + 8x + y + 5$.

This really isn't any more difficult than before, as long as you remember only to collect together things that have the same letters.

Summary

Algebra is no more than using letters to stand for something. Although it sounds complicated, the whole point is to make things easier. We don't want to write down really long expressions involving lots of words, it's easier to use a symbol to stand for something. Similarly we don't want to write something complicated like $5x + 7x - 2x$ when this is just the same as the much simpler $10x$. Algebra is trying to make our life easier, not harder!

Exercises

1 Evaluate the following expressions given the values $a = 4$, $b = 2$.

Example: $3a + b$

Solution: This is the same as $3 \times a + b$. We have $a = 4$ and $b = 2$, so this is the same as $3 \times 4 + 2 = 14$.

(a) $a + b$ (b) $a - b$ (c) ab (d) $\frac{a}{b}$

(e) $b - a$ (f) $2ab$ (g) $a + ab$ (h) $\frac{2b}{a}$

2 Evaluate the following expressions given the values $a = 3$, $b = 4$.

Example: $7a - ab$

Solution: This is the same as $7 \times a - a \times b$. We have $a = 3$ and $b = 4$, so this is the same as $7 \times 3 - 3 \times 4 = 21 - 12$ (remember BODMAS!) $= 9$.

(a) $2a + 4b$ (b) $6a - 5b$ (c) $ab + 2a$ (d) $\frac{ab + b}{a + b - 3}$

(e) $ab - 2a$ (f) $1 - 2ab$ (g) $\frac{7a + b}{ab - 7}$

(h) Why can't you work out $\frac{b - 1}{a - 3}$?

3 Simplify the following expressions.

Example: $2x + 7y + 3x - 4y$

Solution: This is the same as $(2x + 3x) + (7y - 4y)$ which is $5x + 3y$.

(a) $5x + 8x$ (b) $2x + 5x - 9x$ (c) $-3x + 5x - x$

(d) $2x + 5y + 3x + 4y$ (e) $5x + 6y - 2x + y$ (f) $4x + y - 5x - 2y$

(g) $x - 3y - 7x + 3y$ (h) $\frac{x}{5} + \frac{9x}{10} - \frac{4x}{5}$ (i) $\frac{3x}{4} + \frac{5x}{2} + 1$

(j) $3xy + 2 + 5xy - 2y$ (k) $xy + xy + xz - xz$ (l) $1 + x + xy + 1$

4 Answer the following 'real-life' type questions.

Example: If calendars are £4 and diaries are £3 in a shop, write down an expression for the total cost of someone buying in this shop, defining any symbols you use. How much does 5 calendars and 6 diaries cost?

Solution: If c is the number of calendars bought, and d is the number of diaries, then the cost is $4c + 3d$. If you buy 5 calendars and 6 diaries, then $c = 5$ and $d = 6$, so the cost is $4 \times 5 + 3 \times 6 = 20 + 18 =$ £38.

(a) In a sale, CDs are £4 and DVDs are £7.

(i) Write down an expression for the total cost of a purchase in this sale, defining any symbols you use.

(ii) I buy 5 CDs and 2 DVDs. My friend buys 2CDs and 4 DVDs. Work out how much we each spend – which of us spends the most money?

(b) A delivery company charges a flat £10 fee for all orders, and then a further £5 for each package in the order.

(i) Write down a formula for the total cost of an order, defining any symbols you use.

(ii) How much does it cost to send 1 package?

(iii) How much does it cost to send 6 packages?

(iv) How much does it cost to send 100 packages?

9 Brackets in algebra

Using brackets in algebraic expressions

Here we will continue our study of algebra and show how algebraic expressions use brackets in exactly the same way as 'normal numbers' do – this is what you'd expect, since letters in algebra basically just stand for numbers!

Key topics

- → Expanding brackets
- → Simplifying expressions
- → Multiplying out brackets
- → Factorisation

Key terms

algebra and brackets expanding brackets factorising

→ Expanding brackets

Remember from the BODMAS law we introduced in Chapter 1, that things in brackets are always done first, so for example $3 \times (2 + 4)$ is the same as $3 \times 6 = 18$, working out the bit in brackets $(2 + 4)$ first.

Now, note the following: if you multiply both things in the bracket by 3 and add them up, you get $3 \times 2 + 3 \times 4$ which is $6 + 12 = 18$ – the same answer.

So $3 \times (2 + 4)$ is the same as $3 \times 2 + 3 \times 4$.

Does this always work? Check the following are true:

- $4 \times (3 + 2)$ is the same as $4 \times 3 + 4 \times 2$.
- $6 \times (1 + 5)$ is the same as $6 \times 1 + 6 \times 5$.
- $2 \times (3 + 7)$ is the same as $2 \times 3 + 2 \times 7$.

It works when there's a minus inside the brackets too:

- $4 \times (3 - 2)$ is the same as $4 \times 3 - 4 \times 2$.
- $6 \times (1 - 5)$ is the same as $6 \times 1 - 6 \times 5$.
- $2 \times (3 - 7)$ is the same as $2 \times 3 - 2 \times 7$.

In general, this always works. Technically, the word for this is *distributive* but you don't need to know that – but you should know the phrase *expanding the brackets* to describe this process.

Going back to our idea of using symbols to 'stand for something', we can say that for any numbers x, y and z, we have

$$x(y + z) = xy + xz$$

(remember that we don't normally write the multiplication sign $\times$ in algebraic expressions).

→ Expressions involving brackets

Now you know this, you can use this to simplify more complicated expressions. For example, what is $3(x + y) + 2(x - y)$?

Use the rules that we just talked about: $3(x + y)$ is the same as $3x + 3y$. And $2(x - y)$ is the same as $2x - 2y$. So in total, this expression is the same as $3x + 3y + 2x - 2y$. Now if you collect together like terms, as before, then you get $5x + y$.

So $3(x + y) + 2(x - y)$ is the same as $5x + y$, which is much simpler!

Be careful with minus signs. What is $3(x + y) - 2(x - y)$?

Well, $3(x + y)$ is the same as $3x + 3y$, and $2(x - y)$ is the same as $2x - 2y$, just as before. So you have to work out $3x + 3y - (2x - 2y)$ which is the same as $3x + 3y - 2x + 2y$, which now easily simplifies to $x + 5y$.

Don't forget the minus signs when you expand brackets!

The hardest part about doing this was the negative numbers. Don't assume in maths that once you've done a topic, you'll never need it again. For example, you will use negative numbers in many other topics (they are just numbers after all) and they often cause problems – so, regularly go back and quickly revise previous chapters to be sure everything stays in your head.

→ Brackets times brackets

We can use what we developed above to expand even more complicated expressions. What is, for example, $(2 + 3) \times (4 + 5)$? If you work this out, it's 5×9 which gives the answer 45.

Note that if you work out $2 \times 4 + 2 \times 5 + 3 \times 4 + 3 \times 5$ you also get the answer 45 (check this yourself!) Why is this going to be true?

Well, if you used our rules from above about expanding brackets, note that $(2 + 3) \times (4 + 5)$ is the same as $(2 + 3) \times 4 + (2 + 3) \times 5$ (you multiply each bit of the $(4 + 5)$ by what is outside). Because it doesn't matter which order we do multiplication in, this is the same as $4 \times (2 + 3) + 5 \times (2 + 3)$ and we can use our 'expanding brackets' rules again, to get $4 \times 2 + 4 \times 3 + 5 \times 2 + 5 \times 3$, which is the same as $2 \times 4 + 2 \times 5 + 3 \times 4 + 3 \times 5$, just changing the order a little.

What we've done, is basically take every combination of a number from the first bracket and a number from the second bracket:

- The first term is 2×4, which is the first number in $(2 + 3)$ multiplied by the first number in $(4 + 5)$.
- The second term is 2×5, which is the first number in $(2 + 3)$ multiplied by the second number in $(4 + 5)$.
- The third term is 3×4, which is the second number in $(2 + 3)$ multiplied by the first number in $(4 + 5)$.
- The fourth term is 3×5, which is the second number in $(2 + 3)$ multiplied by the second number in $(4 + 5)$.

(Remember that a *term* is one of the things you add up in an expression).

This is illustrated like this:

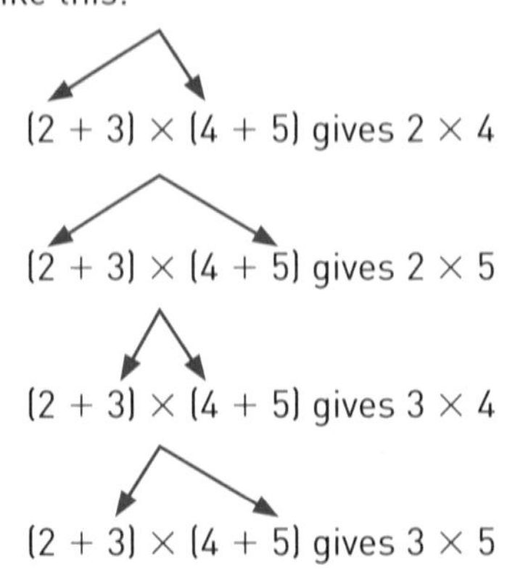

So in total you have:

$$2 \times 4 + 2 \times 5 + 3 \times 4 + 3 \times 5 = 45$$

It works in general, for any numbers, and also then for algebraic expressions.

For example, what is $(x + y)(w + z)$? Taking our combinations just like above, this is the same as $xw + xz + yw + yz$.

Again be careful with minus signs: what is $(x - y)(w - z)$? Again taking combinations but being careful we get all the signs right, this is $xw - xz - yw + yz$.

If you have the same letter in both brackets you can end up with something like $x \times x$. Remember that this is written as x^2. So something like $(x + y)(x + z)$ is the same as $x^2 + xz + yx + yz$ – see below:

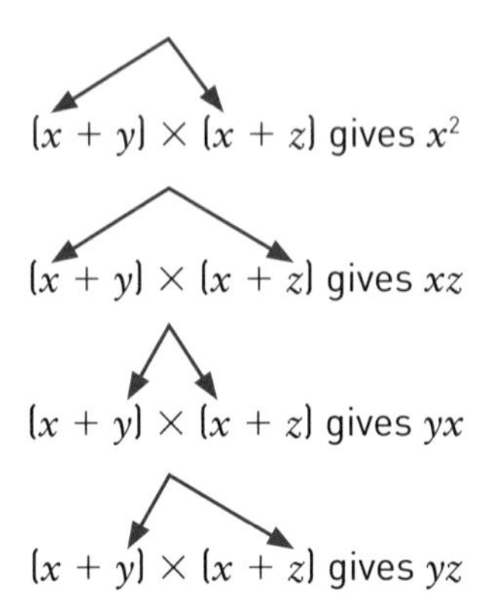

So in total you have:

$$x^2 + xz + yx + yz$$

It works just the same way if there are numbers in front of the symbols. For example, what is $(x + 2y)(3x + z)$?

The first term is $x \times 3x$ which we would normally write as $3x^2$ ($x \times 3x$ is the same as $x \times 3 \times x$, which is the same as $3 \times x \times x$ since the order of multiplication doesn't matter, which we can write as $3x^2$ ignoring the multiplication signs).

Similarly the second term is xz, the third term is $6xy$ (it is conventional to write symbols in alphabetical order but if you write $6yx$ that is fine) and the fourth term is $2yz$.

Hence the final answer is $3x^2 + xz + 6xy + 2yz$.

Note that we always write algebraic terms with the number first then the letters (often in alphabetical order) – so for example we would write $6xy$ rather than $y6x$ – it looks much nicer and makes it easier to follow!

Also, remember to collect terms together. If you do $(x + y)(x + y)$ then you get $x^2 + xy + yx + y^2$. Now, xy is just the same as yx, so these two terms can be collected together and you get $x^2 + 2xy + y^2$.

An interesting one to note is $(x + y)(x - y)$. If you expand this, you get $x^2 - xy + yx - y^2$. Since xy is the same as yx, the two middle terms just cancel to leave you with just $x^2 - y^2$. For example, $(5 + 3)(5 - 3)$ is the same as $5^2 - 3^2 = 25 - 9 = 16$ (check this is right!)

→ Factorisation

Here you've seen how you can expand brackets, so for example we worked out that $x(y + z)$ is the same as $xy + xz$

The reverse process, so given $xy + xz$ and saying that this is the same as $x(y + z)$, is called *factorisation*.

All you have to do is look for something that is in every term of our expression. For example, to do it with numbers first, consider the expression

$$3 \times 4 + 3 \times 5$$

There is a '3' in each term, so we 'bring this out to the front'. What this means is to do the 'multiplying out the brackets' in reverse.

The 3 is in every term, so we write it first, then inside the brackets we fill in all the rest of the terms, but getting rid of the multiplying by 3 in each case. That means we write the 3 first, then inside the brackets write $4 + 5$ (what is left from $3 \times 4 + 3 \times 5$ when we get rid of the multiplying by 3 in each term).

Hence this expression is $3 \times (4 + 5)$. Check that this is the same.

It works exactly the same for symbols too. What is $yz + yx$? Following exactly our rules, we note that there is a y in both terms. 'Bring it out to the front' and you are left with $y(z + x)$ as you would expect.

Don't forget it doesn't matter what order you do multiplying in, and you can 'bring out' any term.

For example, what is $xyz + 3y$? The only thing in both terms is y. 'Bring it out' and you are left with $y(xz + 3)$ (getting rid of the y from the xyz term leaves xz and similarly, getting rid of it from the $3y$ term leaves just 3).

What about something like $6x^2 + 8xy$? You have to be careful here.

The numbers aren't the same, but notice that 2 divides into them both. Also, remember that x^2 is the same as $x \times x$. So, you *could* write this expression as $2 \times 3 \times x \times x + 2 \times 4 \times x \times y$.

Now you can see that there is a $2 \times x$ in both terms, so you can 'bring this out' and you get $(2 \times x)(3 \times x + 4 \times y)$ which you would then write as $2x(3x + 4y)$

After a bit of practice you will be able to do this in one go, but it does take some practice.

smart tip

When doing algebra, it's often a good idea to see if your answers seem right by checking with some actual values. For example, suppose you had to expand $x(y + z)$. You should get $xy + xz$, which you can check by choosing some random values of x, y and z, say $x = 2$, $y = 3$, $z = 4$. Then $x(y + z) = 2 \times (3 + 4) = 2 \times 7 = 14$, and also $xy + xz = 2 \times 3 + 2 \times 4 = 6 + 8 = 14$, the same thing, so you can be confident you got it right! If you didn't get the same thing (say you wrongly wrote $xy + z$ as the answer, then you get $2 \times 3 + 4 = 6 + 4 = 10$) then you know you've got it wrong. Checking your answers like this can really help give you reassurance that you are doing the right thing.

Summary

Using normal numbers, we need to use brackets sometimes. Since algebra is just using symbols to represent numbers, then we need brackets here too, and we'd expect them to follow exactly the same rules as they would with normal numbers. We aren't creating anything magical and new, just doing with the symbols the same as we do with normal numbers!

Exercises

1 Verify the following are true.

Example: $2 \times (3 + 5) = 2 \times 3 + 2 \times 5$

Solution: $2 \times (3 + 5) = 2 \times 8 = 16$. Also, $2 \times 3 + 2 \times 5 = 6 + 10 = 16$, so the two things are equal.

(a) $3 \times (4 + 2) = 3 \times 4 + 3 \times 2$

(b) $2 \times (1 + 1) = 2 \times 1 + 2 \times 1$

(c) $3 \times (4 - 2) = 3 \times 4 - 3 \times 2$

(d) $2 \times (3 - 4) = 2 \times 3 - 2 \times 4$

(e) $-2 \times (3 - 2) = (-2) \times 3 + (-2) \times (-2)$

(f) $2 \times (1 - 1) = 2 \times 1 - 2 \times 1$

2 Expand the brackets in the following algebraic expressions.

Example: $x(y + 3z)$

Solution: Multiply each term by x, so you get the answer of $xy + 3xz$.

(a) $x(y + z)$

(b) $a(b + c)$

(c) $r(s - t)$

(d) $2(x + y)$ (e) $4(x - y)$ (f) $2(2x + 3y)$

(g) $2x(y + z)$ (h) $3x(x - y)$ (i) $xy(x + y)$

(j) $x(3 + y)$ (k) $z(x - 2)$ (l) $2y(x - z)$

3 Expand the brackets in the following algebraic expressions.

Example: $4(x + 2y) + 2(3x - y)$

Solution: This is the same as $4x + 8y + 6x - 2y$ which simplifies to $10x + 6y$.

(a) $2(x + y) + 3(x + y)$ (b) $3(x - 2y) + 2(x + 3y)$

(c) $2(x + y) - 3(x - y)$ (d) $5(2x - 3y) - 2(3x - 6y)$

(e) $x(x + y) + y(2x + y)$ (f) $x(x + y) - y(x - y)$

4 Multiply out the following brackets:

Example: $(x + 2y)(3x + z)$

Solution: Working out the four terms in order as done in the examples above (first term multiplied by first term, etc), you get $3x^2 + xz + 6xy + 2yz$.

Note: remember to collect together like terms where possible.

(a) $(t + u)(r + s)$ (b) $(x + y)(z - w)$ (c) $(x + 2y)(2z + 3w)$

(d) $(x + 2y)(2x + y)$ (e) $(x - 2y)(2x + y)$ (f) $(2x + z)(2x - z)$

5 Factorise the following expressions, checking your final answer.

Example: $2xy + 4xz$

Solution: The two terms have a 2 and an x in common, so 'bringing this out to the front', we get $2x(y + 2z)$. Checking by multiplying out, this is indeed the same as $2xy + 4xz$ so we were right.

(a) $2x + 2y$ (b) $3xy + 2xz$ (c) $3xz + 6xy$

(d) $xyz + xyw$ (e) $10xy + 5y$ (f) $x^2y + xy^2$

10 Solving linear equations

Using algebra to solve linear equations

If I said to you 'I am thinking of a number. If you add 1 to it, you get the answer 5. What is my number?' you'd probably not have too much trouble giving me the answer 4. What's gone in your mind is 'OK, I need one less than 5, which is 4.' All we are going to do in this section is formalise this.

Key topics

→ Linear equations
→ Techniques for solving equations

Key terms

linear equation addition subtraction multiplication division rearranging solving

smart tip

This is a classic example of a subject which looks scary but is actually just modelling what is going on in your mind. If I ask you to give me a number that when I add 5 to it you get 9, your brain will work out that I need 5 less than 9, and give me the answer 4. All we are doing is going through the same process you are performing in your mind, you just don't realise it! Keep this in mind all the way through this algebra section.

→ Linear equations

Suppose you were given the equation $x + 1 = 5$ – this represents our problem mentioned above.

To 'solve' this equation means to work out what values of x make this equation true – alternatively, for what values of x is the value on the left-hand side $(x + 1)$ the same as the value on the right-hand side (5)?

The left-hand side (or just left) refers to what is before the equals sign, and the right-hand side (or just right) refers to what is after the equals sign. Remember that the equals sign says that the left and the right are the same thing.

In this case, the value $x = 4$ is the solution. Evaluating the left-hand side gives $4 + 1 = 5$ which is the same as the right-hand side. So the answer is 4.

That wasn't so hard to work out. But for more complex equations, the answer is not immediately obvious, so you need some general techniques to solve it.

Note that in all the work we will do, we will only consider *linear* equations – that is, those where only x appears, never powers of x like x^2 etc, although we'll see these later in the book (of course, we could use another symbol rather than x).

In what follows, we are always trying to 'rearrange' the equation so that it looks like '$x =$ *something*' where the 'something' doesn't have any xs in it (again of course, we may have a different symbol from x).

In order to achieve this, we are going to try and move the terms around to get it into this form. The informal thing to remember is that to move something from one side to the other, you do its opposite to both sides. What does this mean? We'll illustrate with a series of examples.

→ Adding and subtracting

The opposite of adding is subtracting.

Take the simple equation $x + 3 = 7$. We want $x =$ *something* but we have $x + 3 =$ *something*, so we want to get rid of that $+ 3$ on the left-hand side. The opposite of adding is subtracting. So the opposite of adding 3 is subtracting 3.

You can view this in two ways. One way is to say 'do the same thing to both sides'. If you have two things that are equal, and you do exactly the same thing to each of them, then they still remain equal at the end. In this case we can subtract 3 from both sides:

$$x + 3 - 3 = 7 - 3.$$

and the 3 and -3 cancel on the left, the $7 - 3$ on the right becomes 4, and we are just left with $x = 4$.

In practice, though, we wouldn't probably write this step down, and instead we'd view it as to get rid of the $+ 3$ on the left, it becomes its opposite $- 3$ on the other side.

So $x + 3 = 7$ is the same as $x = 7 - 3$ ('getting rid of' the $+ 3$ by making it its opposite $- 3$ on the other side). You can almost visualise the equals sign as a mirror – to move something from one side to the other, you make it its opposite on the other side.

Similarly, what about $x - 2 = 5$? The opposite of subtracting 2 is adding 2. So to move the $- 2$ to the right, it becomes its opposite $+ 2$, and you get $x = 5 + 2 = 7$.

→ Multiplying and dividing

The opposite of multiplying is dividing.

Consider the equation $2y = 10$.

Remember that $2y$ is short for $2 \times y$ (which is the same as $y \times 2$). The opposite of multiplying by 2 is dividing by 2, so to get rid of the 'multiplying by 2' on the left-hand side, we divide the right-hand side by 2. We get $y = \frac{10}{2} = 5$.

Similarly, the opposite of dividing is multiplying. Consider the equation $\frac{z}{2} = 5$.

To get rid of the dividing by 2 on the left-hand side, we do its opposite, so multiply by 2. We get $z = 5 \times 2 = 10$.

→ Collecting like terms and swapping sides

There are two other techniques you may need to use. One is collecting like terms, just like we did before. If you have the equation $3x + x = 8$ then you collect together the xs on the left-hand side to get the equation $4x = 8$. This is then just a normal equation like above – to get rid of the multiplying by 4 on the left, we divide by 4 on the right, so we get $x = \frac{8}{4} = 2$. So the solution is $x = 2$.

Also, remember that an equation just says one thing is equal to another. It doesn't matter which way round we write the two sides:

$2 + 3 = 5$ says the same thing as $5 = 2 + 3$. This is usually used if we end up with the xs on the right-hand side, when we want them on the left. So, for example, take the equation $3 = x$. This is the same as $x = 3$ (just swapping the two sides over), which gives the solution in the right form.

These basic techniques are all you need to solve any linear equation. Just keep applying them until you get x = *something* (or whatever symbol you are using if not x)

Example 1

Solve the equation $6a - 7 = 2a + 5$

To solve an equation like this, we go step-by-step, applying one of our basic techniques at each step.

- First, we want as on the left-hand side, so we don't want that $2a$ on the right-hand side. To get rid of a positive $2a$ on the right, we subtract $2a$ from the left, giving us the equation $6a - 7 - 2a = 5$. (If it's not clear why we subtract $2a$, think of the right-hand side as $5 + 2a$ instead.)
- Now collect like terms: we have a $6a$ and a $-2a$ on the left-hand side, which is $4a$ in total. So we have the equation $4a - 7 = 5$.
- Now we want to get rid of the -7 on the left. The opposite of subtracting 7 is adding 7, so add 7 to the right-hand side to get $4a = 5 + 7$ and so we have $4a = 12$
- Now get rid of the multiplying by 4 on the right by dividing the left by 4: we get $a = \frac{12}{4} = 3$

So the solution to the equation is $a = 3$. You can check that this solution works by putting $a = 3$ into both sides of the original equation and checking they give the same answer.

The left-hand side is $6a - 7$. Putting $a = 3$ gives $6 \times 3 - 7 = 11$

The right-hand side is $2a + 5$. Putting $a = 3$ gives $2 \times 3 + 5 = 11$

So both sides are the same – we did get the right solution!

Example 2

Solve the equation $\frac{18}{x} = 3$

Again, we will just apply some of the basic techniques.

- We have a dividing by x which isn't nice, we want an x on its own.

So let's get rid of the dividing by x. The opposite of dividing by x is multiplying by x, so get rid of it from the left by multiplying the right by x. We get the equation $18 = 3x$.

- Now we have the xs on the right and we want them on the left. So apply the swapping technique to swap the sides over, giving us $3x = 18$
- Now to get rid of the multiplying by 3, we divide by 3, giving us $x = \frac{18}{3} = 6$

And so we have the answer $x = 6$. Again, check this in the original equation: $\frac{18}{6} = 3$ and so we did get the right answer.

Remember – just apply the basic techniques to get to x (or whatever symbol you are using) = **something. Always check your final answer.**

smart tip

You should always check your answer if you have to do this in a test or in general work. It takes very little time to go back to the original equation and check to see if your answer is right. If it is, you can be sure you got it right and you gain confidence – if not, you know you made a mistake and should go back to look at it.

Summary

This topic is just about solving problems. It's so common in daily life to have to work something out. Just keep in mind that all we're doing when moving things from one side to the other is exactly the same as what is going on in your mind! If I asked you for a number that when I add 3 to it, I get 7, you'd be able to give me the answer 4. Your mind has worked out that you need a number 3 less than 7, so you do $7 - 3 = 4$.

This is all algebra is – it's only doing what you do naturally in your mind anyway!

Exercises

1 Solve the following equations for x.

Example: $x - 5 = 9$

Solution: Remove the -5 on the left by doing its opposite $+5$ on the right, so you get $x = 9 + 5 = 14$.

(a) $x + 3 = 7$ (b) $x + 4 = 1$ (c) $x - 3 = 4$ (d) $x - 4 = -2$

(e) $2x = 12$ (f) $7x = 3$ (g) $\frac{x}{4} = 3$ (h) $\frac{x}{2} = \frac{1}{4}$

2 Solve the following equations for y, checking your answer.

Example: $4y + 5 = 2y + 11$

Solution: Move the $2y$ from the right to the left, you get $4y + 5 - 2y = 11$. Collect together the ys and you get $2y + 5 = 11$. Move the $+5$ to the right and you get $2y = 11 - 5$ and so $2y = 6$. Finally get rid of the multiplying by 2 by dividing by 2 on the other side to get $y = \frac{6}{2} = 3$. Check your answer in the original equation $4y + 5 = 2y + 11$: if you take $y = 3$ then both sides are 17 so you had it right.

(a) $5 = y + 1$ (b) $2y + 4y = 12$ (c) $5y - 2 = 2y + 4$

(d) $6y + 4 = 2y$ (e) $14 - 4y = 2y + 2$ (f) $y = 3y - 1$

(g) $\frac{15}{y} = 3$ (h) $\frac{3}{y} = -3$ (i) $y + 1 = 1 + 2y$

3 Work out my number in the following questions:

Example: If you subtract 5 from my number, you get 10.

Solution: This is the equation $x - 5 = 10$ which solves as $x = 10 + 5 = 15$ and so the number was 15.

(a) If you add 3 to my number, you get 7

(b) If you multiply my number by 3, you get 18

(c) If you divide my number by 2, you get 3

(d) If you multiply my number by 6, and then add 3, you get 27

(e) If you add 5 to my number, and then multiply by 2, you get 20

(f) If you multiply my number by 4 you get -2

(g) If you multiply my number by 6 and add 2, you get the same answer as if you added 1 and then multiplied by 2.

11 Transposition and algebraic fractions

Rearranging expressions using basic rules of algebra

You saw before how to solve equations by moving things around from one side to another, essentially mimicking how your mind works. Here we're going to do the same sort of thing but use symbols – remember that the symbols just stand for numbers, so that we should be able to do the same of thing with them as we do with normal numbers.

Key topics

→ Transposition
→ Algebraic fractions

Key terms
Transposition rearranging equations subject algebraic fraction

→ Transposition

Often we will have more than one variable (unknown) in an equation. Go back to our idea of salaries and bonus payments that we discussed in Chapter 8. If m is your usual monthly payment, b is your Christmas bonus payment, and d is your December pay, then we have the formula $d = m + b$. If you are told what m and b are, you can work out d.

Now look at a slightly different question. If I tell you what my December pay and usual monthly pay is, what is my bonus? What I want to do is work out b – I want to rearrange this formula so it says $b =$ *something*. This is called making b the ***subject*** of the formula.

You follow exactly the same approach as we did with solving equations – use the basic techniques to get the equation into the right form.

- We start with the equation $d = m + b$ and want $b =$ *something*
- All the bs are on the right and we want them on the left, so swap the sides round: $m + b = d$
- Now we want to get rid of the m on the left. To do this, subtract m from both sides, just as we did earlier. You get $b = d - m$.

So this is our formula with b as the subject. So for example, if my December pay (d) was £1800 and my usual monthly pay (m) is £1600, I can work out my bonus as $d - m = 1800 - 1600 = 200$. So my bonus was £200.

smart tip

Remember that all we are doing is formalising what is actually going on in your head. If I had told you that my December salary was £1800 and my normal salary was £1600, you'd probably have been able to tell me that my bonus was £200 by subtracting the two values. Algebra isn't anything 'weird', it's just modelling the way your mind actually works! Keep this in mind as you work through maths – it's not another language, it's just expressing what your mind already does!

Remember, just apply the basic techniques!

Example

Make s the subject of the formula $x + ys = 3x + t$

This looks complicated, but just apply the basic rules to make it $s =$ *something*.

- We want to get rid of the x on the left. So subtract x from both sides to get $ys = 3x + t - x$
- Collect like terms: We have a $3x$ and a $-x$ on the right which gives $2x$. So we have $ys = 2x + t$
- Now we want to get rid of the multiplying by y on the left, so divide both sides by y and we get $s = \dfrac{2x + t}{y}$

So if you were told what x, y, and t were, then you could work out s. For example, if $x = 3$, $t = 8$, $y = 2$, then $s = \dfrac{2 \times 3 + 8}{2} = \dfrac{14}{2} = 7$.

Moving formulae around like this is called *transposition*.

You should choose which things to do to make things as easy as possible. For example, take the equation $z = y - x$ where you want to make x the subject.

You *could* swap the sides first, giving $y - x = z$ but the problem is that if you then take the y away, you get $-x = z - y$ which is quite hard to deal with (you've got $-x$ rather than x).

It would be better to move the $-x$ to the left, so you get $z + x = y$, which then gives us $x = y - z$.

When you are dividing, remember that the opposite of division is multiplication. So, for example, if you have $\frac{x}{y} = z$ and you want x to be the subject, then to get rid of the dividing by y, you multiply by y on the other side: $x = yz$.

My general advice would be to get rid of negatives and fractions as soon as you can, since they are much harder conceptually than 'nice' positives, additions and multiplications! The following example is more complicated but hopefully you can follow the steps.

Example

Make x the subject of the expression $\frac{y}{x} - z = w$.

This takes a few steps.

First move the z to the right-hand side – you get $\frac{y}{x} = w + z$

Now, we don't like dividing, so let's get rid of the dividing by x. To do this, we multiply by x on the other side, so we get $y = x(w + z)$

Now we need to get rid of the multiplying by $(w + z)$, so we get

$$\frac{y}{w + z} = x$$

Finally, we just swap the sides round, so we get $x = \frac{y}{w + z}$

I'm not in any way pretending that was really easy – it's not. But it does come with practice – experience helps you make the right choice as to what move to make.

Remember:

- **Adding and subtracting are opposites.**
- **Multiplying and dividing are opposites.**
- **You can swap the sides, and collect like terms.**

Your aim is to get *'subject = something'*

→ Algebraic fractions

We saw earlier how to deal with an expression like $\frac{x}{2} + \frac{x}{4}$ – we made the denominators the same, and added up in the usual way:

$$\frac{x}{2} + \frac{x}{4} = \frac{2x}{4} + \frac{x}{4} = \frac{3x}{4}$$

You can still apply exactly the same ideas when the fractions get more complicated and have symbols rather than numbers on the bottom.

For example, what is $\frac{r}{s} + \frac{t}{u}$ the same as, as a single fraction?

Well, if you multiply the top and the bottom of the first term by u, you get $\frac{ru}{su}$. Similarly, if you multiply the top and bottom of the second term by s, you get $\frac{st}{su}$. Now, the denominators (bottom bits) are the same, and we have to work out $\frac{ru}{su} + \frac{st}{su}$, which adding up just like normal, gives us $\frac{ru + st}{su}$, and we have successfully turned the original expression into a single fraction.

The most important thing in all of this is to realise that you aren't actually doing anything different from what you've done before with fractions.

As another example, what is $\frac{2}{y} + \frac{1}{xy}$?

Multiply the top and bottom of the first term by x and this is the same as

$\frac{2x}{xy} + \frac{1}{xy}$ which then adds up to $\frac{2x + 1}{xy}$.

You can deal with subtraction in exactly the same way, and multiplication and division are just as easy as with numbers:

$$\frac{a}{b} \times \frac{c}{d} = \frac{ac}{bd}$$

(multiplying the top bits together, and the bottom bits together)

$$\frac{a}{b} \div \frac{c}{d} = \frac{a}{b} \times \frac{d}{c} = \frac{ad}{bc}$$

(using the normal division rule of turning the second fraction upside down and multiplying).

smart tip

Check your answers in algebraic expressions by putting real numbers in and checking it works. If you do so, then you gain the realisation that algebra is just using symbols to stand for numbers, which you do know how to manipulate. Algebra is not something different, it's just the same rules of maths with a symbol instead of a number!

Summary

Take a very simple equation, such as one which relates your take-home pay to your actual salary and your tax deduction. Different people would have different questions. One might know their salary and tax, and want to know what they take home. Another might know their salary and what they take home, but might want to know their tax. Yet another might want to know their salary if they know their take-home pay and their tax.

This is all basically the same equation. If we use the rules discussed in this topic, then we (or a computer) can rearrange the basic equation into the right form so that the different people identified above can all get an answer to their problem.

Also remember that we are just expressing what goes on in our minds. In your head, you can work out these sorts of transpositions – algebra is nothing different, just writing down on paper what's going on in your head when you try to work things out.

The more you realise that maths is really just representing your actual thoughts, the less you will be scared of it and the more confident and able you will become with the subject!

Exercises

There are various ways to get to the right answer in each of these – as long as you get the right answer, that's fine!

1 Make s the subject of the following formulae, and evaluate s when $r = 2, t = 4$.

Example: $t = 2s + r$

Solution: Swap the sides to make $2s + r = t$. Move the r to the other side (so it becomes $-r$ on the other side) to get $2s = t - r$. Then remove the multiplying by 2 by dividing by 2 on the other side to get $s = \frac{t - r}{2}$.

When $r = 2$ and $t = 4$ this gives the answer $\frac{4 - 2}{2} = \frac{2}{2} = 1$.

(a) $r = s - 2t$ (b) $t = r - s$ (c) $r = t - 2s$

(d) $t = \frac{r + s}{2}$ (e) $r = 3 + ts$ (f) $t = \frac{rt + 1}{s}$

2 Make x the subject of the given formulae.

Example: Make x the subject of $3x - 4y = s + t$

Solution: Move the $4y$ to the other side to make $3x = s + t + 4y$ and then divide by 3 to make $x = \frac{s + t + 4y}{3}$.

(a) $x - y = z$ (b) $x + y = z - w$ (c) $z = y + x$

(d) $z = y - x$ (e) $2x + y = z$ (f) $3z = y + 2x$

(g) $xy = z$ (h) $xy = z + 1$ (i) $\frac{x}{y} = z$

(j) $\frac{y}{x} = z$ (k) $\frac{x}{y} - z = w$ (l) $\frac{z}{2x} + 3y = 4w$

3 Express the following as single fractions in their lowest terms.

Example: $\frac{3}{x} + \frac{z}{y}$

Solution: Make the denominators the same – this is the same as

$$\frac{3y}{xy} + \frac{xz}{xy} = \frac{3y + xz}{xy}.$$

(a) $\frac{2x}{7} + \frac{3x}{7}$ (b) $\frac{x}{3} + \frac{x}{4}$ (c) $\frac{7x}{11} - \frac{3x}{11}$

(d) $\frac{5x}{8} - \frac{3x}{8}$ (e) $\frac{1}{x} + \frac{1}{y}$ (f) $\frac{4}{x} + \frac{3}{y}$

(g) $\frac{x}{y} \times \frac{z}{w}$ (h) $\frac{x}{y} \times \frac{y}{z}$ (i) $\frac{8}{x} \div \frac{2}{x}$

12 Solving simultaneous equations

Solving two equations at the same time

Instead of trying to solve one equation like we have done before, let's try to really extend ourselves and solve two equations at the same time.

Key topics

- → Simultaneous equations
- → Six basic steps to solving simultaneous equations

Key terms

simultaneous equations

→ Simultaneous equations - introduction

Let's suppose I gave you an equation like $2x + 4y = 14$. Can you come up with a solution to this (so values of x and y that make it work? After you tried for a while, you might come up with the solution $x = 3$ and $y = 2$. This works, if you put in $x = 3$ and $y = 2$ you get $2 \times 3 + 4 \times 2 = 6 + 8 = 14$ which is the right answer.

But it's not the only solution. Try $x = 1$ and $y = 3$. Or $x = 7$ and $y = 0$. Or $x = -1$ and $y = 4$ ... and many, many more. Equations like this, where you have two things to find, will always have infinitely many possible answers.

Here's another equation: $5x + 3y = 21$. Can you come up with any answers for this? Again, there are infinitely many – you could have $x = 0$ and $y = 7$ for example, or maybe $x = -3$ and $y = 12$. But another solution is $x = 3$ and $y = 2$. Note that this is the same as what we found above for the previous equation – the answers $x = 3$ and $y = 2$ make both equations true.

Can you come up with any other answers that are true for both of them? That is called solving *simultaneous* equations (solving two equations 'at the same time'). As much as you try, you won't be able to find anything else that works. There is one, and only one, answer that works for both of them – namely $x = 3$ and $y = 2$.

Equations like this are called *linear* – you need know nothing more than that simply means there aren't any powers, so nothing like x^2 or y^3, just x and y.

For any two linear equations (at least for any that you come across for now), there will be one, and only one, solution that works for BOTH of them.

→ Six steps for solving simultaneous equations

The question remains as to how to find this solution. In the above, we sort of guessed it and chanced upon $x = 3$ and $y = 2$. We'll go through a systematic technique that you can use to solve any pair of simultaneous equations, using the example above and then another example. If you follow these steps exactly, you will be able to solve any pair of simultaneous equations that you are likely to meet.

The first step

Write down the two equations.

Number them (1) and (2). In our case it is:

$$(1)\ 2x + 4y = 14$$
$$(2)\ 5x + 3y = 21$$

The second step

Make the *x*s the same.

Multiply each equation by something so that in the end, the number of xs are exactly the same. Let's show by example:

If you multiply the equation (1) by 5 you get this:

$$10x + 20y = 70$$

You multiply *every term* by 5: 5 lots of $2x$ are $10x$; 5 lots of $4y$ are $20y$; 5 lots of 14 are 70. So you get the equation above.

Now if you multiply the second equation (2) by 2 (so multiply every term by 2) you get:

$$10x + 6y = 42$$

The important part is that we made the number of xs the same – both are $10x$.

Have a look at our two equations now, renumbering them (3) and (4):

$$(3)\ \ 10x + 20y = 70$$
$$(4)\ \ 10x + 6y = 42$$

The third step

Subtract the two equations from each other.

This means to go through, first with the xs, then with the ys, then with the numbers, and take the second one away from the first. Again we'll show by illustration:

Do (3) − (4).

First the xs: $10x - 10x$ gives nothing at all – that's the whole point about making the xs the same. So the xs have gone away.

Now the ys: $20y - 6y = 14y$

Now the numbers: $70 - 42 = 28$.

So we are left with $14y = 28$ (the xs have gone away)

The fourth step

Solve for *y*.

Now you can just solve the previous equation and work out what y is.

We have $14y = 28$, and hence $y = \frac{28}{14} = 2$.

The fifth step

Choose your favourite of the original equations and put in the value for *y*.

In this case we have $y = 2$. *It doesn't matter which equation you choose* – let's choose $2x + 4y = 14$. If we put $y = 2$ then this gives us:

$$2x + 4 \times 2 = 14, \text{ so}$$

$$2x + 8 = 14, \text{ so}$$
$$2x = 6, \text{ and so}$$
$$x = 3.$$

Hence we have worked out the answer: $x = 3$ and $y = 2$.

Check for yourself that you get exactly the same answer if you choose the other equation ($5x + 3y = 21$) instead.

The sixth step

Check your answer.

Note that we can easily check we got it right – put these values back into the original equations and make sure we get the right answer.

We had $2x + 4y = 14$ and $5x + 3y = 21$. In both cases, if you substitute in $x = 3$ and $y = 2$ you do indeed get the correct answer (make sure you do check this yourself) – so we did get it right!

smart tip

Sadly, for something like this, there isn't much you can do apart from just learn the basic rules. But once you've learnt them, then you can solve any simultaneous equations you will see. The best way to learn is probably by practice. The more exercises you do and practise with, the more the rules will stick in your mind – when you have to learn rules, apply them practically and you'll find them easier to learn.

→ Another example

Let's do another example to illustrate. We can do exactly the same as before, the only slightly tricky thing is to be careful with negatives, as this one has minus signs in it. Suppose we have the two equations

$$4x + 2y = 22$$
$$7x - 4y = 1$$

We will just go through our six steps as we did before.

The first step

Write down the two equations.

Number them (1) and (2). In our case it is:

$$(1)\ 4x + 2y = 22$$
$$(2)\ 7x - 4y = 1$$

The second step

Make the *x*s the same.

Multiply each equation by something so that in the end, the number of xs are exactly the same.

If you multiply the equation (1) by 7 you get this:

$$28x + 14y = 154$$

Remember, you multiply *every term* by 7: 7 lots of $4x$ are $28x$; 7 lots of $2y$ are $14y$; 7 lots of 22 are 154. So you get the equation above.

Now if you multiply the second equation (2) by 4 (so multiply every term by 4) you get:

$$28x - 16y = 4$$

Remember that the important part is that we made the number of xs the same – both are 28.

Our two equations now, renumbering them (3) and (4), are:

$$(3)\ 28x + 14y = 154$$

$$(4)\ 28x - 16y = 4$$

The third step

Subtract the two equations from each other.

This means to go through, first with the xs, then with the ys, then with the numbers, and take the second one away from the first. Again we'll show by illustration:

Do (3) $-$ (4).

First the xs: $28x - 28x$ gives nothing at all – that's the whole point about making the xs the same. So the xs have gone away.

Now the ys: $14y - (-16y) = 30y$ (be very careful with the minus signs)

Now the numbers: $154 - 4 = 150$.

So we are left with $30y = 150$ (the xs have gone away)

The fourth step

Solve for *y*.

Take the equation we had before $30y = 150$.

Hence $y = \frac{150}{30} = 5$.

The fifth step

Choose your favourite of the original equations and put in the value for *y*.

In this case we have $y = 5$. *Remember that it doesn't matter which equation you choose* – let's choose $4x + 2y = 22$. If we put $y = 5$ then this gives us:

$$4x + 2 \times 5 = 22, \text{ so}$$
$$4x + 10 = 22, \text{ so}$$
$$4x = 12, \text{ and so}$$
$$x = 3.$$

Hence we have worked out the answer: $x = 3$ and $y = 5$.

The sixth step

Check your answer.

Put these values back into the original equations and make sure we get the right answer. If you do put $x = 3$ and $y = 5$ into both equations

$$4x + 2y = 22$$
$$7x - 4y = 1$$

you do get the right answers, so we did it right!

→ Real-life simultaneous equations

You can solve any simultaneous equations using these basic steps. Sometimes, as in the next example, the problem will be written in words and you have to create the equations first, but you still follow the same steps to solve them.

Example

At a fast-food restaurant, 3 burgers and 5 fries cost £4.82. 4 burgers and 3 fries cost £4.41. How much do burgers cost and how much are fries?

We will just follow our basic six steps:

Step 1

First of all, write this down as a pair of equations. Let b be the cost of a burger and let f be the cost of fries.

The first statement tells us that 3 burgers, and 5 fries, together, make £4.82. So, we have $3b + 5f = 4.82$ (we won't write the pound sign down as it gets confusing in the equations). The second statement tells us that 4 burgers, and 3 fries, together, make £4.41. So, we have $4b + 3f = 4.41$.

Hence our two equations are:

$$(1)\ 3b + 5f = 4.82$$

$$(2)\ 4b + 3f = 4.41$$

Step 2

Now 'make the x's the same'. The only thing is that it isn't x in this example, it's b, but we still do the same thing, so make the bs the same. Multiplying the first equation by 4 (remembering to multiply *every* term by 4) you get $12b + 20f = 19.28$ (you will probably need to use a calculator for that). Multiplying the second equation by 3 (again remembering to multiply *every* term by 3) you get $12b + 9f = 13.23$. So now our equations are:

$$(1)\ 12b + 20f = 19.28$$

$$(2)\ 12b + 9f = 13.23$$

Step 3

Next, 'take the equations away from each other. $12b - 12b$ is nothing at all, that's the point. $20f - 9f = 11f$ and $19.28 - 13.23$ is 6.05. So now we have:

$$11f = 6.05$$

Step 4

Calculate f – we have $f = \frac{6.05}{11} = 0.55$ (using a calculator).

Step 5

Pick your favourite of the original equations. Remember, it doesn't matter which you choose – let's just say we choose $3b + 5f = 4.82$.

Then setting $f = 0.55$ as we just worked out, we have:

$$3b + 5 \times 0.55 = 4.82, \text{ and so}$$
$$3b + 2.75 = 4.82, \text{ and so}$$
$$3b = 4.82 - 2.75 = 2.07$$
$$\text{and so } b = \frac{2.07}{3} = 0.69$$

Hence we worked out the answers to be $b = 0.69$ and $f = 0.55$.

Step 6

You should check for yourself that these answers do actually work in both equations. Note that our original total was given in pounds (£) and so these answers are in pounds too: the cost of a burger is £0.69 (or you could write 69p) and the cost of fries is £0.55 (or you could write 55p).

This question may have seemed more complicated than our first question, but that's only because it had decimals in it and needed some use of a calculator to do the division. We still followed the set of six basic steps.

Remember – if you follow these basic steps then you can solve virtually any simultaneous equations you will come across.

→ Extra optional section

This section is extra and entirely optional – if you find it any way confusing, ignore it and concentrate only on being able to apply the basic rules above.

On some occasions, you might find that there's a quicker way to solve your equations. For example, take the pair of equations:

$$3x + 2y = 13$$
$$5x - 2y = 11$$

If instead of making the xs the same, what happens if you just add up the two equations? Adding up the xs, you get $3x + 5x$ which is $8x$. Adding up the ys you get $2y - 2y$ which is nothing at all. And adding up the 13 and 11 you get 24.

So you get $8x = 24$ (the ys have gone away) and so $x = 3$.

Hence, putting $x = 3$ into either of the original equations, you get $y = 2$.

This was quicker than making the xs the same, but it only worked because the number of ys in the top equation was the negative of the number of ys in the bottom equation, so by adding them up, the ys went away.

If you can spot such shortcuts, feel free to use them, of course, but the basic rules above will work for any equations that you are likely to see.

Also, it is actually possible to have simultaneous equations you can't solve. For example, what about the equations $3x + 2y = 5$ and $3x + 2y = 6$? Whatever the values of x and y, you can't possibly come up with $3x + 2y$ both being 5 and 6 at the same time, that would just say that 5 is the same as 6, and you know 5 isn't 6!

Also, what about $3x + 2y = 5$ and $6x + 4y = 10$? If you multiply every term in the first equation by 2, you get $6x + 4y = 10$ which is exactly the same as the second equation. So these two equations are the same. So it's just the same as having one equation, which you know has infinitely many possible answers.

smart tip

Focus on the basics. In this chapter, if you learn the basic rules then you can solve pretty much any problem you're likely to meet. If you're really comfortable and can handle some extra topics (like we discussed above) then that's great and can make things easier – but in your studies, focus on being able to do the basic things first before moving on to anything harder. You are bound to be good at some things and not so good at others – expand your knowledge in things you are good at, and at least be able to do the basic in things you are not so good at.

Summary

Here we're coming up with a mathematical way to solve what is actually quite a hard problem. If you took the problem with the burgers and the fries, say, then to solve that in your head would have taken you ages, having to keep checking possibilities until

eventually you probably hit on the answer. But it is a 'real-life' problem, the sort of which might really need to be solved.

The rules to follow here are each individually fairly straightforward. I know it's not easy to learn them all – but it comes with practice. Whilst it's probably not true to say that 'practice makes perfect', it's probably true to say that 'practice makes better' – practice as much as you can with such things and they become easier over time – it's the same with pretty much anything!

Having completed this chapter and hence the algebra section, hopefully you now have a much better appreciation that algebra really is just modelling normal situations and you can approach the subject with more confidence knowing that fact!

Exercises

1 Solve the following simultaneous equations.

Example: $4x + 3y = 22$ and $3x - 2y = 8$

Solution:

Step 1 – write the equations down:

$$4x + 3y = 22$$
$$3x - 2y = 8$$

Step 2 – 'make the xs the same', multiply the first equation by 3 and the second equation by 4.

$$12x + 9y = 66$$
$$12x - 8y = 32$$

Step 3 – subtract the equations. Be careful with the minus sign: $9y - (-8y) = 17y$.

$$\text{You get } 17y = 34$$

Step 4– Solve for y

$$\text{You get } y = \frac{34}{17} = 2$$

Step 5 – Choose your favourite equation and solve for x.

Choosing $4x + 3y = 22$ (you can choose the other one if you like), and filling in $y = 2$, you get:

$$4x + 3 \times 2 = 22$$
$$4x + 6 = 22$$
$$4x = 22 - 6$$
$$4x = 16$$
$$x = 4$$

Hence $x = 4$ and $y = 2$

Step 6 – Check your answer. $x = 4$ and $y = 2$ works in both equations $4x + 3y = 22$ and $3x - 2y = 8$

So the answer is $x = 4$ and $y = 2$.

(a) $3x + 2y = 16$
$5x + 3y = 26$

(b) $4x + 3y = 18$
$5x - 2y = 11$

(c) $4x + 3y = 15$
$6x - 5y = 13$

(d) $5x + y = 11$
$3x + 2y = 1$

(e) $x + 4y = -5$
$3x - 5y = 2$

(f) $3x + 5y = 8$
$x - 5y = -4$

2 Solve the following questions expressed 'in words'.

Example: 3 burgers and 5 fries costs £5.47; 5 burgers and 4 fries costs £6.95. How much are burgers and how much are fries?

Solution: This gives the two equations (where b = cost of a burger and f = cost of fries):

$$3b + 5f = 5.47$$
$$5b + 4f = 6.95$$

Make the bs the same by multiplying the first equation by 5 and the second equation by 3:

$$15b + 25f = 27.35$$
$$15b + 12f = 20.85$$

Subtract the equations:

$$13f = 6.5$$

Hence $f = 0.5$, and then you can work out $b = 0.99$ from either equation.

Therefore burgers are 99p (£0.99) and fries are 50p (£0.50).

(a) A sale has posters all at the same price, and CDs all at the same price. Alice buys 5 posters and 3 CDs and spends £30. Bob buys 2 posters and 4 CDs and spends £26. How much are posters, and how much are CDs?

(b) Xavier buys 5 pears and 3 apples for £4.81. Yolande buys 3 pears and 7 apples for £5.59. How much are pears and how much are apples?

(c) If you add together two numbers you get 82, and if you subtract them from each other you get 6. Write this as a pair of simultaneous equations, and hence work out what the two numbers are.

DATA

13 Presentation of data

Using various types of graphs and charts to illustrate data visually

In this chapter, we are going to investigate some basic elements of data presentation and statistics – we will look at ways in which collections of data can be presented in an appealing visual form, or calculations can be performed, to give us an indication of what the overall data is telling us.

Key topics

- → Analysing data
- → Plotting graphs
- → Drawing charts

Key terms

data line graph scatter graph correlation bar chart pie chart

→ What is this all about?

Statistics is all about collecting data and presenting it in a useful form. There are different types of data that could be collected. For example, you might conduct a survey on which party people plan to vote for at the next election, or perhaps observe the population of a beehive at daily intervals. Maybe you would measure the radioactivity of a chemical substance every second, or collect data about the marks in an examination. There are many different types of data.

For most collections of data, it's not particularly useful to just have the raw list of data. For a start, you might well have a huge collection of data, and anyway you really want to draw general conclusions about the overall data. For example, when election polls are given on the news, you don't have a list of every single person surveyed, you just get given an overall figure (e.g. '42% of people voted this way', rather than a list of all people and their votes, and you have to work out the percentage yourself – that would take ages!).

You might see data given in the form of graphs (e.g. the inflation rate over the last 2 years), pie charts and bar charts (e.g. voting intentions) and various other ways.

Note, by the way (although we won't investigate this here), that there are issues in how you collect the data. When surveying people for voting intentions, for example, you want the data to be representative of the whole population, so you want to choose a good sample of people. Collecting the data by standing outside a political headquarters and asking people who go in who they plan to vote for is hardly going to be representative.

But it gets much deeper than this. For an election poll, what would be wrong with conducting a poll by email, say? The problem is that a certain class of people (e.g. young professionals) are more likely to have email addresses than others (e.g. the older generation) and so you aren't getting a representative sample of the whole population.

Similarly with scientific data, you need to be sure, for example, that the act of taking the measurement hasn't altered or disrupted what you are trying to measure.

smart tip

Next time you read a newspaper or watch the news, count how many times you see data presented, be it in the form of a graph, a picture, a list of percentages, or other ways. Realising that the maths is being used here to get across the 'results' of the data analysis helps you appreciate its importance.

→ Line graphs

When you have a series of measurements taken over time, they are often represented as a line graph. This involves plotting all the points on a graph, and connecting them together from one point to the next. You often see this sort of graph in the news – for instance showing the interest rate over the last few months or years. For example, here are the UK inflation rates for an 18-month period in 2005–06:

Jan 05	Feb 05	Mar 05	Apr 05	May 05	Jun 05	Jul 05	Aug 05	Sep 05
3.2	3.2	3.2	3.2	2.9	2.9	2.9	2.8	2.7
Oct 05	Nov 05	Dec 05	Jan 06	Feb 06	Mar 06	Apr 06	May 06	Jun 06
2.5	2.4	2.2	2.4	2.4	2.4	2.6	3.0	3.3

To plot these on a graph, you need to create two *axes* (the plural of *axis*), one going across (which will represent our months) and one going up (which will represent the inflation rate). For each month, go up to the appropriate inflation rate and mark the point with a cross or a solid point or some other clear way. For example, to plot the first value, go to the label of Jan 05, and then go up until you reach 3.2, marking it in some way.

Note that you should label the axes clearly as to what corresponds to what.

Once you have filled all the values in, then connect them together with a series of straight lines as in the following.

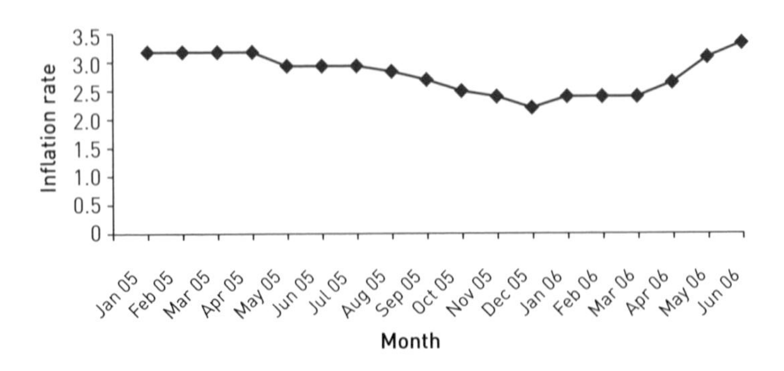

It's more obvious from this graph what is the overall trend, rather than from just looking at a long list of numbers. It's obvious the inflation rate was falling, and then started rising in around December 2005.

For any set of data which is taken over a period of time at regular intervals, a line graph is usually a good way to show the overall trend.

→ Scatter graphs

A scatter graph is when you simply plot every point on a graph without making any attempt to connect them together. It can be used for data when both the horizontal and vertical axes of the graph can take a range of values (not necessarily in order of time, just a random collection of data). If the data is 'scattered' widely then there appears to be no *correlation* (relationship) between the two things. If they form something close to a line, then there is a good correlation. For example, the following data might have been obtained, recording people's height (in metres) and their weight (in kg).

Height (m)	1.75	2.01	1.64	1.51	1.72	1.81	1.65	1.92	1.83	1.87
Weight (kg)	80	109	72	60	81	90	73	101	93	99

On a scatter graph, this could be plotted as:

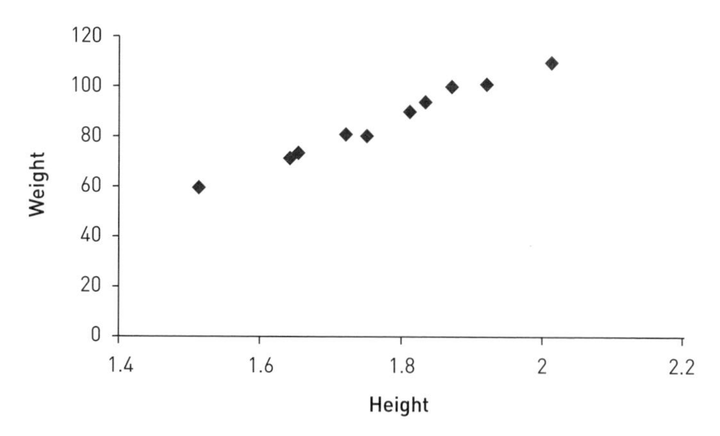

(Note the scale – why start at 1.4 not 0?)

There appears to be a strong correlation between height and weight – they seem to fall pretty well in a line. This is what you might expect – you would expect taller people to weigh more.

Now here is data analysing the same people, with the number of pairs of shoes they own:

Height (m)	1.75	2.01	1.64	1.51	1.72	1.81	1.65	1.92	1.83	1.87
No. of shoes	5	3	14	8	2	10	9	20	1	7

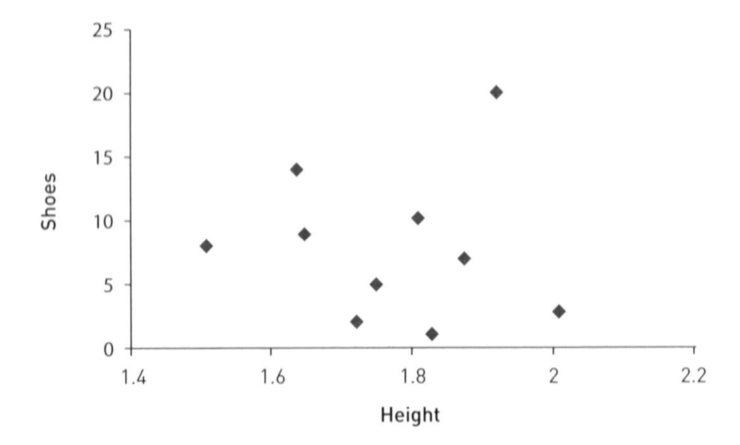

This time there doesn't appear to be a connection – the points are scattered everywhere! Again, this is probably what you would expect – why would a person's height relate to how many shoes they have?

Also note that there would be no point drawing a line between the points here – we use a line graph when we have a series of measurements taken over time, but for a general set of measurements where there is no time flow to them, we just use a scatter graph.

→ Bar charts

When the data you get can be classified into a small number of 'categories', then you can often express it usefully as a bar chart.

For example, suppose customers at a store were asked to rate their shopping experience as 'good', 'average', or 'poor'. The results were as follows:

Good	32 people
Average	78 people
Poor	6 people

To plot this as a graph, label the horizontal axis with 'good', 'average', and 'poor', and then draw 'bars' for each one to represent the appropriate number, like this:

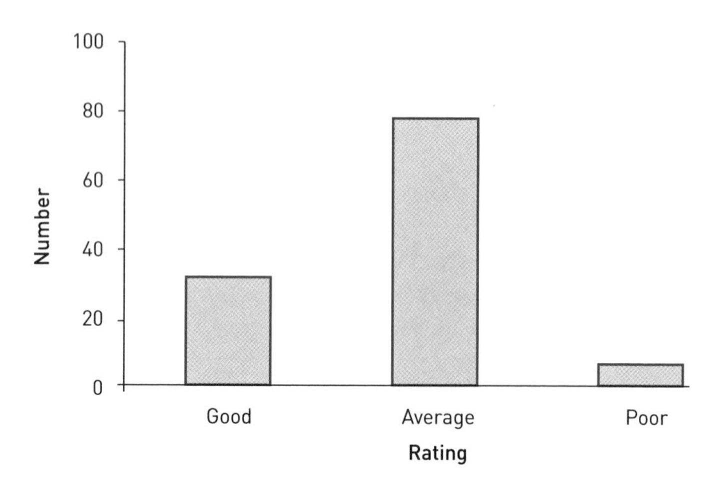

→ Pie charts

A pie chart is a way of representing proportions. For example, 50 people were surveyed and asked which football team they support. The results were as follows:

Man Utd	23
Chelsea	11
Liverpool	8
Arsenal	6
Others	2

To draw these figures on a pie chart, we are going to divide a circle into segments corresponding to the options – the larger the segment, the more people chose that option.

A circle is made up of 360 degrees (written as 360°). You should know how to use a protractor to measure an angle – if you don't know, then ask someone to show you how to use one!

smart tip

Do get hold of a basic maths set. You can buy them really cheaply from standard stationery stores – they can come in useful even at times you don't expect; there are lots of times you want a ruler to draw a straight line or something!

The proportion of people who opted for Man Utd is $\frac{23}{50}$. A circle has 360° in it. The technique to follow is to convert this proportion so it is out of 360 – that is the number of degrees that the 'Man Utd' segment should use. Do this:

- Work out the proportion as a decimal on your calculator. In this case we get $\frac{23}{50} = 0.46$
- Multiply this number by 360 on your calculator – you get 165.6
- So the Man Utd segment should be 165.6°

Similarly, you will find that the Chelsea segment should be 79.2°, the Liverpool segment should be 57.6°, the Arsenal segment should be 43.2° and the Others segment should be 14.4°.

Draw a circle and draw a vertical radius (a line from the centre to the edge) like this:

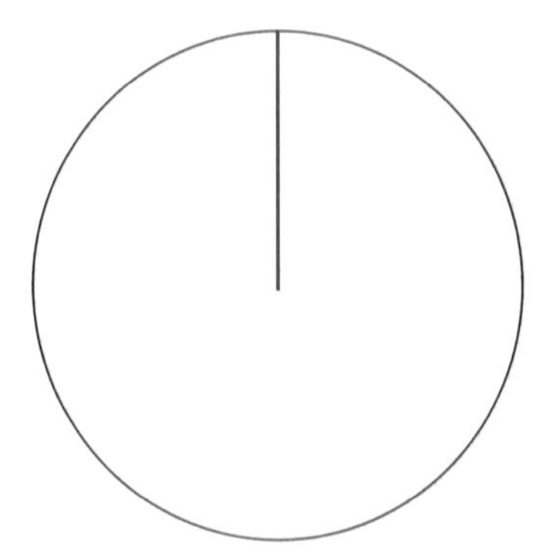

Now, *carefully* with a protractor, draw the Man Utd segment – so it should form an angle of 165.6°. You get something like this:

Continue like this for each segment. Label each segment clearly. When presented as part of a formal document, using different colours for each segment is also recommended.

Sometimes in pie charts, we would also include the numbers, or, more commonly, the percentage of people making each choice, as below (check the percentages are correct). So, the final pie chart in our example looks like this:

Summary

It is vital to present data in an easy-to-understand way. When people see the results of a by-election, for example, they don't just want to see the actual number of votes cast for each candidate, they want the statistics broken down into a nice visual format that makes the result jump out at them. Not everyone can immediately see what a series of inflation rate figures mean, but a picture shows the trend.

People don't want to understand long lists of figures, they want a simple picture summarising them. 'A picture paints a thousand words' is a well-known phrase; in the same vein, maybe a picture paints a thousand figures!

Exercises

These questions are similar to what has been discussed above, so no examples are supplied here.

1 (a) Present the following data showing the number of visitors to a website each day as a line graph:

Day	1	2	3	4	5	6	7	8	9	10
Visitors	46	52	53	60	61	24	25	30	32	37

(b) Present the following data showing the interest rate of a bank over 12 months as a line graph:

Month	Jan	Feb	Mar	Apr	May	Jun	Jul	Aug	Sep	Oct	Nov	Dec
Rate	10.1	10.5	11.8	13.7	12.9	12.9	13.1	12.0	11.5	10.9	11.4	11.9

2 The following data shows the average temperature in successive weeks, and the corresponding figures for sales of ice-cream and CDs in a store. Plot the data on two scatter graphs (one for temperature and ice-creams, and one for temperature and CDs). Does there appear to be a correlation in either case?

Temperature	20	25	32	21	28	23	18	14	12	8
Ice creams	120	220	300	115	250	170	120	40	45	10
CDs	90	60	50	120	130	100	50	30	60	90

3 (a) In a survey of a website, the percentage of people rating the quality of the site in various categories was:

Excellent 16%
Good 43%
Average 22%
Poor 13%
Very poor 6%

Plot this information as a bar graph and as a pie chart.

(b) In a class election, the 40 votes were cast as:

John 6 votes
Sarah 22 votes
Ahmed 9 votes
Nina 3 votes

Plot this information as a bar graph and as a pie chart.

14 Measures of location and dispersion

Working out 'averages' and a measure of the spread of a set of data

As well as presenting our data in a nice form, it is also useful to have some basic figures 'representative' of the data. For example, it's useful to know the 'average' mark in a class test, and get some indication of how the marks were spread – was there a huge range of marks, or were they concentrated in a small range?

Key topics

- → Location and dispersion
- → Averages and ranges
- → Standard deviation

Key terms

mean median mode range variance standard deviation

→ Measures of location

Measures of location refer to figures that measure, in some way, the 'middle', 'average', or 'central location' of a collection of data. There are three basic measures of location which you need to know how to work out.

Mean

The most common measure of location is the *mean* (or formally, *arithmetic mean*). This is the same as what you probably usually refer to as the *average* of a set of values.

To work out the mean:

- **Add up all the values.**
- **Divide by the number of values.**

For example, to work out the mean of the numbers 12, 18, 16, 14:

- Add up the values: $12 + 18 + 16 + 14 = 60$.
- Divide by the number of values. There are four values, so work out $\frac{60}{4} = 15$.

The mean, or average, is 15.

Similarly, here is a list of marks in a class test:

56 80 94 18 40 45 71 62 50 59 63 6 98 26 69

To work out the mean, add them all up: you get 837. Then divide by the number of values (there are 15 marks here) so the mean is $\frac{837}{15} = 55.8$ (I wouldn't expect you to do that in your head, you can use a calculator!). So the mean, or average mark, is 55.8.

Note that the mean isn't always the best way to represent an average. With salaries, for example, while most people earn salaries in a small range of around £10,000 to £50,000, say, there are a few people with exceptionally high salaries of many million pounds. These very high salaries push the average up, meaning that most people earn less than the 'average'.

Median

The *median* of a set of values can be thought of as the 'central value' where, roughly speaking, half of the values are below it, and half of the values are above it.

To work out the median:

- **Write down the values in ascending order (smallest to largest).**
- **If there are an odd number of values, pick the one in the middle.**
- **If there are an even number of values, pick the two in the middle and take their mean.**

So, for example, to work out the median of the figures 4, 9, 100, 16, 1, first of all write them in ascending order: 1, 4, 9, 16, 100. There are an odd number of values (5), so pick the figure in the middle of the list – this is 9. So the median is 9.

Similarly, work out the median of 16, 4, 64, 49, 9, 36 (*Aside: there is a connection between these numbers, what is it?*) Writing them in ascending order: 4, 9, 16, 36, 49, 64. This time there are an even number of values (6) so there is no single 'middle' point. Hence, pick the two middle values (16 and 36) and take their mean, which is $\frac{16+36}{2} = \frac{52}{2} = 26$. Hence the median is 26.

Medians are commonly used with things like salaries – half of the population have a salary less than the median and half have a salary more than the median.

Example

The following is a list of salaries (measured in thousands of pounds):

19 25 14 17 35 1000 43 11 31 22 15

To work out the median salary, write the numbers in ascending order:

11 14 15 17 19 22 25 31 35 43 1000

There are 11 values, which is an odd number, so pick the one in the middle, which is 22. So the median salary is £22,000. Half of the people have a salary less than this, and half have a salary more than this.

Note that the mean salary (add up the figures and divide by 11) is £112,000. Which of these do you think is the better indicator of the 'average' salary?

Mode

The *mode* of a set of values is the most commonly occurring value. So to work it out:

Just count which value appears the most times.

For example, the mode of the numbers 3, 1, 8, 2, 3, 3, 5, 1, 1, 3, 2, 7 is 3, since 3 appears the most times (four times).

Note that this definition (unlike mean and median) makes sense for values which aren't numbers. For example, the list of students in a small class is:

Ahmed, John, Sarah, Nina, John, Michael, Emma, John, Ahmed, Yang, Sarah.

In this class the most common name is John (there are three Johns, and only two Sarahs and Ahmeds), so the mode is John.

On some occasions there may be more than one mode. For example, what is the mode of the numbers 1, 1, 1, 2, 2, 3, 3, 3, 4? The numbers 1 and 3 both appear three times and so both are the most common. Hence the mode is 1 and 3.

These three measures of location are all you need to know at this stage, although there are other measures used in practice (in particular, there are many types of mean such as geometric mean and harmonic mean, which you could research if you wanted to).

smart tip

Make up your own examples here. Find out the mean, median and mode of the marks you have obtained in a series of tests or exams, your weekly shopping bill, or anything else you can think of! Make maths real by applying it to real life.

→ Measures of dispersion

As well as wanting some way to measure the 'average' (mean), 'middle' (median), or 'typical' (mode) value, it can also be of interest to know how much spread or variety there is in our data. Are most of our values concentrated in a small range, or are they massively spread out over a wide range?

Range

The simplest measure of the range of values is simply to work out the spread from the smallest to largest value. So just:

Subtract the smallest value from the largest value.

For example, the range of 16, 19, 18, 20, 17, 18, 16 is $20 - 16 = 4$, whereas the range of 4, 9, 8, 2, 7, 6, 3, 10000, 5, 8, 8, 3 is $10000 - 2 = 9998$.

This is a very weak way of measuring the spread. In the second example above, the 10000 seems to be an anomaly (very different from the other values) – maybe an error occurred in our reading there? Also, it takes no account of the values apart from the smallest and largest, so there is no measure of how well spread they are.

Variance and standard deviation

The *variance* is a measure of how much the values differ from the mean. To work it out, do the following:

- **Calculate the mean**
- **For each value, work out the difference between the value and the mean**

- **Square this difference (multiply it by itself)**
- **Add up all the squared differences**
- **Divide by the number of values.**

So, for example, consider the following values:

10 6 8 5 4 9

The mean of these values is 7 (the sum is 42 and there are 6 values, so the mean is $\frac{42}{6} = 7$)

So, for each value, work out the difference between the value and the mean, and then square it. So for example, in the first column, the difference between the value (10) and the mean (7) is 3, which when squared gives 9.

Values	10	6	8	5	4	9
Difference	3	−1	1	−2	−3	2
Diff squared	9	1	1	4	9	4

Note that square numbers are always positive – this is one of the reasons we square the differences, to avoid any problems with minus signs. So 4 is exactly the same distance away from the mean as 10 is (both 3 away) – by squaring the differences we get the same 'difference squared' value for them both.

Now add up the squared differences to get $9 + 1 + 1 + 4 + 9 + 4 = 28$, and divide by 6 to get 4.67 (to 2dp). So the variance is 4.67 (to 2dp).

The larger the variance, the more the values are spread.

The *standard deviation* is simply the square root of the variance. So, the standard deviation of this set of values is $\sqrt{4.67} = 2.16$ (using a calculator, to 2dp). Again, the larger the standard deviation, the more the numbers are spread. In our example, the standard deviation is quite small, so the values are not very spread.

Again, there are other measures of dispersion, such as interquartile range and mean dispersion, but you do not need to know about these at this stage.

The word 'average' appears commonly in daily life – average house prices, average speed, etc. This is just meant to summarise something – remember that you are summarising a huge chunk of stuff that you don't need to know (every house price for example!) and giving a 'representative' figure. Often adverts will claim that a certain percentage of people liked their product. You do this sort of 'summarising' very often – again, the maths isn't meant to be scary, it's meant to help!

Summary

Presenting summaries of data is a vital aspect for businesses and people in general. If someone asked you 'what is the public's view of your product?' you wouldn't give them a list of every single person and their opinion, you'd summarise the data, perhaps say that 87% rated it excellent, for example.

When the Government (from the Office of National Statistics) gives the financial situation for a year, say, they don't list every single transaction made, no one could read that in their lifetime! They use measures to summarise the data, so declare an overall growth of 1% (for example).

There is so much data in the world nowadays that it is essential to have summary measures to give a basic indication of the data there – this is where these measures come in!

Exercises

For each of the following sets of data, calculate the:

(a) mean
(b) median
(c) mode
(d) range
(e) variance
(f) standard deviation

Give any decimal answers to 2 decimal places.

Example: Cost of an item in pounds:

3 3 8 1 5

Solution:

(a) Adding them all together gives $3 + 3 + 8 + 1 + 5 = 20$, and there are 5 numbers in the list, so the mean is $\frac{20}{5} = 4$.

(b) Putting them in order gives 1, 3, 3, 5, 8 – as there is an odd number of things here, there is only one middle number 3 so this is the median = 3.

(c) The mode is the most common number = 3 (it appears twice, everything else only appears once)

(d) The range = largest number – smallest number = $8 - 1 = 7$

(e) Variance: use the table like above, remembering that the mean is 4 as you worked out before:

Values	3	3	8	1	5
Difference	−1	−1	4	−3	1
Diff squared	1	1	16	9	1

Add up the difference squared to get $1 + 1 + 16 + 9 + 1 = 28$, and then divide by the number of things (5) to get $\frac{28}{5} = 5.6$, so the range is 5.6

(f) The standard deviation is the square root of the variance, so $\sqrt{5.6} = 2.37$ (2dp)

1 Marks obtained in a class test:

6 9 10 3 7 6 8

2 Ants in a colony:

42 51 25 17 76 19 4 17 42 17

3 Goals scored in football matches:

3 3 0 2 8 2

15 Straight lines

Linear data and calculating the equation of a straight line

Often we have a collection of data which fits naturally into a pattern. In this chapter we will look at data which follows a *linear* (straight-line) pattern and work out how to express such data mathematically.

Key topics

→ Straight-line graphs
→ Gradients and intercepts
→ Equation of a straight line

Key terms
linear data axes x-axis y-axis straight line graphs gradient intercept '$y = mx + c$'

→ Straight lines - an example

Consider the following data, which indicates the monthly telephone bill received by a customer, depending on how much time they have spent talking in the month:

Number of hours	2	4	6	8
Bill in pounds (£s)	14	18	22	26

Is there a pattern to this data? How much would the bill be if the person talked for 10 hours? 20 hours? What about if they didn't make any calls at all?

Let's plot this data on a graph. To draw a graph, you draw two axes like those below, and 'plot' the points of data, drawing little crosses corresponding to the data points – so, for example, we draw a cross at the point corresponding to 2 hours and 14 pounds, then a cross at 4 hours and 18 pounds, etc.

Note that we take the first thing (number of hours in this case) to go horizontally (labelled on the so-called x-axis) and the second thing (the bill) to go vertically, labelled on the so-called y-axis.

You should draw this accurately yourself on graph paper. You should get something like the one below.

smart tip

Graph paper isn't that easy to find and can be expensive – but it's quite easy to download free internet applications that can print out graph paper for you on a standard printer. Whilst studying, take advantage of free things if you are on a limited budget – don't waste your money!

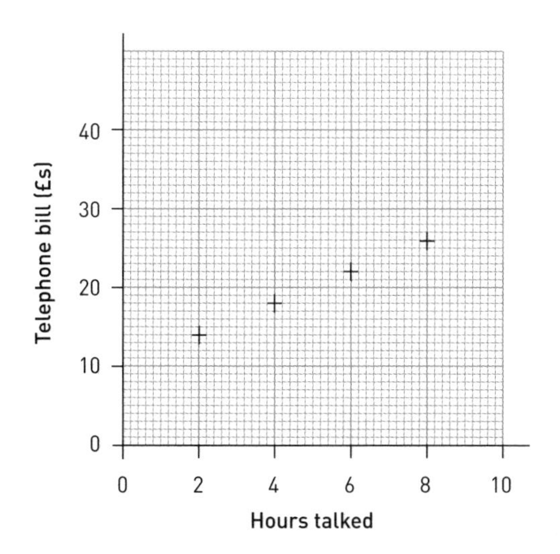

It looks as though these points lie in a straight line. Carefully draw a straight line through all of the points, extending it in both directions. If you have drawn your graph and plotted the points accurately, you should find that the line passes perfectly through all four points, and you should end up with something like the graph below.

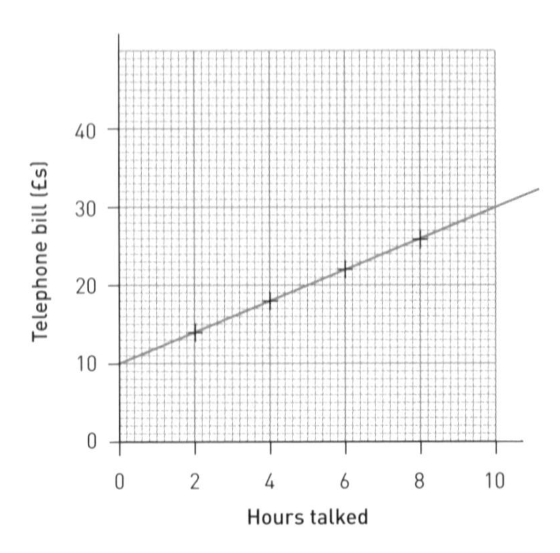

Now, use your graph to answer the questions:

(i) What would be the bill for 10 hours' talking?

(ii) What is the bill if no calls are made?

To answer (i), look at the point on the line corresponding to 10 hours. You should find that it corresponds exactly to a bill of £30. For (ii), look at the point corresponding to 0 hours. This is where the line 'hits' the axis corresponding to the bill. You should find this is £10 exactly. So even if you make no calls at all, you spend £10. This is what you might consider as a 'standing charge' or the charge for line rental.

What about for 20 hours? Or 100 hours? Our graph doesn't go that far. What we need is a formula to help us work the bill out for a given number of hours.

The general equation for a straight line is this:

$$y = mx + c$$

where y represents the quantity on the vertical axis (y-axis), x represents the quantity on the horizontal axis (x-axis), m represents the *gradient* (or *slope*), and c represents the *intercept* (where the line cuts the y-axis).

Remembering that m stands for gradient and c for intercept isn't that easy – there's no particular reason why we use these letters, it's just become standard over time.

So for our graph, y represents the bill, and x represents the hours spent talking. How do we work out m and c? Follow these rules exactly.

To work out the gradient m:

- **Choose *any* two points on the line (x_1, y_1) and (x_2, y_2)**
- **Calculate $m = \frac{y_2 - y_1}{x_2 - x_1}$**

So for our example, pick *any* two points – that is, pick any two bits of data. We will choose the pairs (2, 14) and (4, 18) – that is, the points on the line corresponding to (2 hours talking, £14 bill) and (4 hours talking, £18 bill). We know these are on the line because they are part of the original data.

Calculate $m = \frac{18 - 14}{4 - 2} = \frac{4}{2} = 2$

So our gradient (slope) is $m = 2$.

To work out the intercept c:

- **Pick *any* point on the line (x, y) and use the value m calculated above.**
- **Calculate $c = y - mx$ (this comes from just rearranging (transposing) the formula)**

So for our example, pick *any* point – so any bit of data. We will choose the pair (2, 14) – the point corresponding to (2 hours talking, £14 bill) which is part of our original data.

So we have $x = 2$, $y = 14$, and $m = 2$ from above.

Hence $c = y - mx = 14 - 2 \times 2 = 14 - 4 = 10$

Note that this corresponds to where the line 'hits' the y-axis and matches the £10 bill for no calls made that we worked out from our graph earlier.

Hence the equation of the line representing our data is $y = 2x + 10$

We can use this formula to work out the bill for as many hours as we like.

For 20 hours: $x = 20$ and so $y = 2 \times 20 + 10 = 50$. So the bill is £50 for 20 hours talk.

For 100 hours: $x = 100$ and so $y = 2 \times 100 + 10 = 210$. So the bill is £210 for 100 hours talk.

You can imagine the intercept (10) as the line rental, and the gradient (2) as the 'cost per hour' – line rental is £10 and calls are £2 per hour.

Exercise

Repeat the above calculations choosing different pairs of values when you calculate m and c, and check that you get the same answers.

→ Second example – negative gradients

In the last case, the gradient was a positive number. When the gradient is positive, the line slopes 'up' towards the right – the bigger the gradient, the steeper the slope. When the gradient is negative, the line slopes 'down' towards the right – the bigger negative it is, the more it slopes. When the gradient is zero, the line is absolutely 'flat'.

Remember that a positive gradient slopes up, and a negative gradient slopes down.

Example

Consider the following data which gives the acidity of a chemical compound for different temperatures.

Temperature in °C (x):	-10	10	30	50
Acidity measure (y):	14	8	2	-4

What is the equation of the straight line passing through these points?

To work out the gradient m, pick any two points. To avoid using negative numbers as much as possible, let's pick the pairs (10, 8) and (30, 2), but you can choose any pair you like.

Then the gradient is $m = \dfrac{2-8}{30-10} = \dfrac{-6}{20} = \dfrac{-3}{10} = -0.3$

Now to work out c: pick any point, the simplest one is (10, 8). Then $c = y - mx = 8 - (-0.3 \times 10) = 8 - (-3) = 11$.

So the equation of the line is $y = -0.3x + 11$

Exercise

Plot these points on a graph. You will need negative numbers on your axes so your axes should probably look something like the graph below.

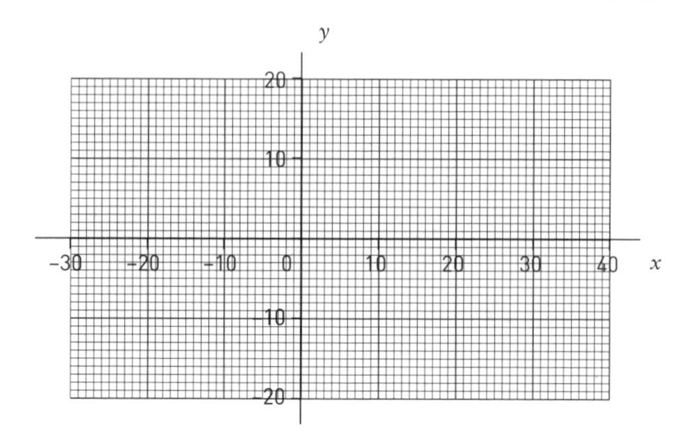

If you plot it accurately, the line should slope downwards (negative gradient) and cut the y-axis at 11 (intercept)

smart tip

This topic is about taking data and drawing a line to 'fit it'. Whilst straight lines and linear data are quite simple, it gives the idea behind more complicated graphs. As humans, we find things quite easy to see 'visually' – keep an eye on how many times you see visual representation of data, and remember that a mathematician was behind it to create the visualisation!

Summary

Lots of data has linear (straight line) graphs like this, for example we talked above about a 'fixed charge' and then a 'charge per hour' – you'll be aware of this if you've booked a plumber or other such worker to come out – there is a 'call-out charge' and then a charge per hour for the work. It is important to have an algebraic formula for it, so that for example the worker can immediately give a quote for 50 hours' work, without having to draw a really big graph and look it up!

It's also important to present data. A straight-line graph makes sense to people, they understand it's going up or down to represent

increases or decreases. Given that the majority of people nowadays, even the older generation, are computer-literate and do want access on their PCs to data in a way they understand, it is important to be able to present data visually.

This topic of course goes much deeper than simple straight-line graphs but it is an important introduction to a deep and complex area.

Exercises

You will need graph paper for this – you can obtain printable graph paper from the internet if you cannot buy it. You may well need a calculator for some of these calculations.

Example: The following data shows the electricity bill for a customer, depending on the number of units they have used.

Number of units	2	4	6	8
Bill in pounds (£s)	13	21	29	37

(a) Draw a graph of these points and draw a straight line through them. Use your graph to give the bill if they used 7 units.

(b) Work out the equation of the line in the form $y = mx + c$

(c) What do the gradient and intercept represent in this example? Check that your equation gives the same answers as in (a) for 7 units.

(d) What would the bill be for 30 units used?

Solution:
(a) Any accurate graph is fine, it should look something like this:

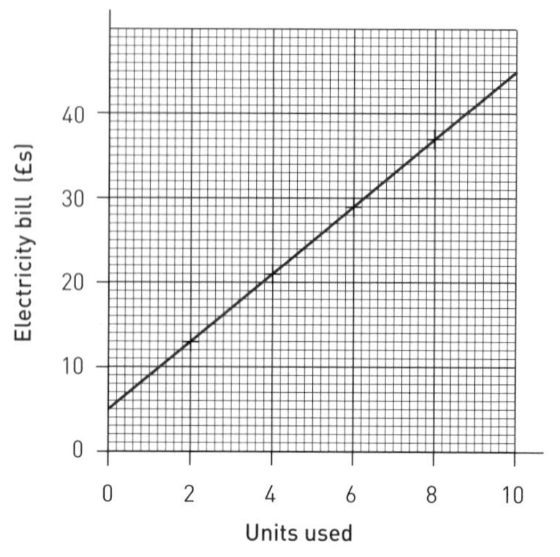

From the graph, 7 units costs £33 (look it up on the line).

(b) Pick any two pairs – say (2,13) and (4,21). Then the gradient $m = \frac{21 - 13}{4 - 2} = \frac{8}{2} = 4$

Pick any pair, say (2,13). The intercept $c = y - mx = 13 - 4 \times 2 = 13 - 8 = 5$

Hence the equation is $y = 4x + 5$.

(c) The gradient represents a 'charge per unit' and the intercept represents a fixed 'standing charge'. Putting $x = 7$ into the equation gives $4 \times 7 + 5 = 33$, the same as in (a) from the graph.

(d) Put $x = 30$ into the equation to get $4 \times 30 + 5 =$ £125.

1 The following data shows the pressure rating in a cabin compared with the outside temperature.

Temperature in °C (x)	−10	0	10	20
Pressure rating (y)	15	20	25	30

(a) Draw a graph of these points and draw a straight line through them. Use your graph to give the pressure rating at temperatures 40°C and −20°C

(b) Work out the equation of the line in the form $y = mx + c$

(c) Check that your equation gives the same answers as in (a) for temperatures 40°C and −20°C.

(d) What would the pressure rating be at temperature 100°C?

2 The following data shows the temperature of an object compared with its altitude.

Altitude in m(x)	1000	2000	3000	4000
Temperature in °C (y)	30	28	26	24

(a) Draw a graph of these points and draw a straight line through them. Use your graph to give the temperature at altitudes 0m and 5000m

(b) Work out the equation of the line in the form $y = mx + c$

(c) Check that your equation gives the same answers as in (a) for altitudes 0m and 5000m.

(d) What would the temperature be at altitude 15000m?

3 A garage charged the following bills for hours worked on cars:

Hours worked	0.5	1.0	1.5	2.0
Bill (£s)	115	130	145	160

(a) Calculate the equation of this line.

(b) What do the gradient and intercept represent in this question?

(c) Use your formula to calculate the cost of 3 hours work.

4 A builder's merchant charges the following for block deliveries:

No. of blocks	100	200	300	500
Cost (£s)	£54	£83	£112	£170

(a) Calculate the equation of the line.

(b) What do the gradient and intercept represent in this question?

(c) Use your formula to calculate the delivery cost of 20 blocks.

(d) Use your formula to calculate the delivery cost of 2000 blocks.

16 An introduction to probability

Working out probabilities - the chances of something happening

A lot of what happens in life is down to chance. Businesses have often failed or succeeded due to lucky or unlucky events, and for every lottery winner there is a gambler who lost everything. Chance and randomness are part of our life, but we can at least predict what 'might' happen. In a horse race, some horses are more likely than others to win, so there are different odds given to them. If you were on the deciding question in Who Wants to be a Millionaire and your 'phone-a-friend' was 90% sure they knew the answer, would you go for their answer?

Key topics

- → Probability
- → Calculating discrete probabilities
- → The probability of multiple events

Key terms

probability events complement disjoint multiple events

→ Idea of probability

Before we define it mathematically, let's think about what a probability 'should' mean. We'll take this entirely informally. Take for example a normal die (if you didn't know, *die* is the singular noun, *dice* is the plural, so it's wrong to say 'a dice'), with six sides, numbered 1, 2, 3, 4, 5 and 6.

Suppose you throw this die randomly. What are the 'chances' you throw a 1? Just think this out from an overview point of view. There are six possibilities – you might get a 1, a 2, a 3, a 4, a 5 or a 6 – and each one is equally likely. So the chances of getting a 1 might be seen as '1 in 6' – basically, 1 out of the 6 equally likely possibilities.

You could write this as a fraction – the probability of throwing a 1 is $\frac{1}{6}$.

What is the chance you score more than 1? Well, out of the six possible outcomes, all of them are more than 1, apart from 1 itself. So if you get 2, 3, 4, 5 or 6, then you have scored more than 1. So five out of the six possibilities mean you score more than 1. So the probability of scoring more than 1 is $\frac{5}{6}$.

smart tip

> This topic can be made fun by illustrating with games – you can roll dice or toss coins or think about the lottery – and maybe then when playing games you can work out the likely chance of you winning. Perhaps you can even create your own games. Making maths 'fun' makes it much more interesting!

→ Discrete probabilities

Following on from the above example, we can generalise the concept of the 'probability' of an *event* (a specific happening) occurring. We will look only at what are called *discrete* probabilities – that is where there are a fixed number of distinct possible outcomes.

The definition of the probability of an event occurring is given by:

$$\text{P(event)} = \frac{\text{(the number of outcomes in which that event can occur)}}{\text{(the total number of possible outcomes)}}$$

It isn't very obvious what this means, so let's illustrate with an example.

Suppose we are picking a random integer (whole number) between 1 and 10. Hence, there are 10 possible outcomes: 1, 2, 3, 4, 5, 6, 7, 8, 9 or 10.

We are going to calculate the probability (chance) of picking the following. (Note that we can give probabilities either as fractions or decimals – but as we discussed before, you should generally only use decimals when they are exact answers (not rounded).)

(i) 5

(ii) A number less than 8

(iii) An even number (divisible by 2)

(iv) An odd number (not divisible by 2)

(v) Anything apart from 5

(vi) At least 1

(vii) 100

- For (i), only one of the ten possible outcomes gives 5, and so

$$\mathbf{P}(\text{picking } 5) = \frac{1}{10} = 0.1$$

- For (ii), there are seven out of the ten possible outcomes which give an answer which is less than 8 (namely 1, 2, 3, 4, 5, 6, and 7), and so

$$\mathbf{P}(\text{picking a number less than } 8) = \frac{7}{10} = 0.7$$

- For (iii), there are five out of the ten possible outcomes which give an answer which is even (namely 2, 4, 6, 8, and 10), and so

$$\mathbf{P}(\text{picking an even number}) = \frac{5}{10} = 0.5$$

- For (iv), there are five out of the ten possible outcomes which give an answer which is odd (namely 1, 3, 5, 7, and 9), and so

$$\mathbf{P}(\text{picking an odd number}) = \frac{5}{10} = 0.5$$

- For (v), nine out of the ten possible outcomes give an answer which is not 5 (namely 1, 2, 3, 4, 6, 7, 8, 9 and 10), and so

$$\mathbf{P}(\text{picking a number other than } 5) = \frac{9}{10} = 0.9$$

- For (vi), all ten of the possible outcomes are at least 1, and so

$$\mathbf{P}(\text{picking a number that is at least } 1) = \frac{10}{10} = 1$$

- Finally for (vii), none of the ten possible outcomes can give 100, it's impossible to obtain as an outcome, and so

$$\mathbf{P}(\text{picking } 100) = \frac{0}{10} = 0$$

Note in the last two examples, that if our desired outcome is certain, we have probability 1, and if the desired outcome is impossible, we have probability 0.

→ Complements

Given an event, the *complement* of that event is essentially 'the complete opposite' of it.

For example, what is the complement (opposite) of 'picking 5' in the above example? It is those outcomes that mean we *don't* pick 5. The outcomes that make these true are as follows:

Picking 5: outcome of 5

Not picking 5: outcome of either 1, 2, 3, 4, 6, 7, 8, 9, 10

Note that every outcome appears once in these lists: either in the first list or the second one.

Also note that the sum of the probabilities (as we did above) is

$$\mathbf{P}(\text{picking 5}) + \mathbf{P}(\text{not picking 5}) = \frac{1}{10} + \frac{9}{10} = 1$$

Similarly, what is the complement of 'picking an even number'? It's the complete opposite: you don't pick an even number, or in other words, you pick an odd number.

Even number: outcomes of 2, 4, 6, 8, 10

Odd number: outcomes of 1, 3, 5, 7, 9

Note, again, that every possible outcome appears once in these lists, either in the first list or the second list. And again note that the sum of the probabilities is 1:

$$\mathbf{P}(\text{even number}) + \mathbf{P}(\text{odd number}) = \frac{5}{10} + \frac{5}{10} = 1$$

This rule is true in general: if you add together the probability of an event, and the probability of its complement, you get 1.

→ Either/or

Sometimes questions are asked where the desired outcome can be expressed in the form 'either one thing, or something else'. For example, what is the probability, when you roll a die, that the outcome is either 1, or greater than 3?

Well, there are six possible outcomes, of which four of them make the desired outcome (that is, 1 (which satisfies the first condition), and 4, 5, 6 (which satisfy the second condition)). So **P**(either 1, or greater than 3) $= \frac{4}{6} = \frac{2}{3}$ (you could write this as a decimal 0.67 to 2dp but it's more accurate to leave it as a fraction).

Now, note the following:

$$\mathbf{P}(\text{outcome of 1}) = \frac{1}{6}$$

(1 is the only one of the six possible outcomes that is OK)

$$\mathbf{P}(\text{outcome of greater than 3}) = \frac{3}{6} = \frac{1}{2}$$

(4, 5, 6 are OK from the six possible outcomes)

Now, if you add up these probabilities, you get $\frac{1}{6} + \frac{1}{2} = \frac{4}{6} = \frac{2}{3}$ which is the same answer as we had above. Does this always work? **No**.

It works only when the two events are *disjoint* – which means they have nothing in common. For example, I roll the die again – what is the probability I either get a number less than 3, or an even number?

Well, the outcomes that make this work are either 1, 2 (numbers less than 3), or 2, 4, 6 (even numbers). So in total, a roll of 1, 2, 4, 6 is OK, and 3 and 5 aren't – four possibilities out of the six. So the probability is **P** either less than 2, or even) $= \frac{4}{6} = \frac{2}{3}$.

$$\text{But } \mathbf{P}(\text{less than 3}) = \frac{2}{6} = \frac{1}{3}$$

$$\mathbf{P}(\text{even}) = \frac{3}{6} = \frac{1}{2}$$

And if you add these up you get $\frac{1}{3} + \frac{1}{2} = \frac{5}{6}$ which is NOT the right answer.

The problem comes because the number 2 is in both lists:

Less than 3: outcomes 1, 2

Even: outcomes 2, 4, 6

You see that 2 appears in both lists, so they are not *disjoint* (they have something in common, namely 2) which means you can't just add up the probabilities.

You can only add up the probabilities when the lists are disjoint. Unless you are completely happy with all of this, I suggest you work out the probability of either/or events just by listing all the outcomes as usual.

→ Multiple events

Suppose you both tossed a coin, and rolled a die. What is the probability that:

(i) the coin is heads and the die rolls 6,

(ii) the coin is tails and the die rolls an even number?

This is somewhat trickier. There are many combinations that give possible outcomes. The coin can either be a head or a tail, and the die can either roll 1, 2, 3, 4, 5 or 6.

In total there are twelve possible outcomes – writing them as pairs (coin toss, die roll) they are:

(head, 1), (head, 2), (head, 3), (head, 4), (head, 5), (head, 6), (tail, 1), (tail, 2), (tail, 3), (tail, 4), (tail, 5), (tail, 6).

So for (i) the only one of these that satisfies our requirement is the pair (head, 6) and so our probability is $\frac{1}{12}$ (one of the twelve possible outcomes).

For (ii) there are three possible outcomes that are OK – (tail, 2), (tail, 4) and (tail, 6). So our probability is $\frac{3}{12}$ (three of the twelve possible outcomes) which is $\frac{1}{4}$.

Now, note the following:

- In (i), the probability of a head is $\frac{1}{2}$ and the probability of rolling a six is $\frac{1}{6}$. If you multiply these together you get $\frac{1}{2} \times \frac{1}{6} = \frac{1}{12}$ which is the probability you just worked out of both events happening together (both a head, and a six).
- In (ii), the probability of a tail is $\frac{1}{2}$ and the probability of rolling an even number is $\frac{3}{6} = \frac{1}{2}$. If you multiply these together you get $\frac{1}{2} \times \frac{1}{2} = \frac{1}{4}$ which is the probability you just worked out of both events happening together (both a tail, and an even number).

This works in general as long *as the two events are unconnected*: **if you multiply the probability of one event by the probability of another, you get the probability of them both happening.**

→ Multiple events – more advanced example

Now consider what happens when you toss two coins. What is the probability that you get:

(i) two heads,

(ii) one head and one tail?

You have to be very careful here. What are the possible outcomes?

It's NOT true to say that there are three outcomes (two heads, two tails, or a head and a tail) and so both probabilities are $\frac{1}{3}$.

If you consider the outcome of the first coin and the second coin separately, after some thought you will find that the only four possible outcomes are:

First coin	Second coin
HEAD	HEAD
HEAD	TAIL
TAIL	HEAD
TAIL	TAIL

So there are four possible outcomes. Hence the answer to the questions is:

(i) the only one of the four possible outcomes that gives us two heads is the first row of the table, so **P**(two heads) $= \frac{1}{4}$

(ii) in both the second and third rows of the table, we have one head and one tail. Hence **P**(a head and a tail) $= \frac{2}{4} = \frac{1}{2}$.

The important thing here is that 'a head and a tail' is interpreted as being the same as 'a tail and a head' and so it's twice as likely to happen as getting two heads.

Experiment for yourself. If you aren't convinced by the above, for example, then take two coins and throw them yourself a good number of times (say 50) and you should find that around half the time you get a head and a tail. Roll a die 50 times and you'll probably get roughly half even and half odd. Maybe you can come up with your own games? Make maths practical and fun and it will be so much easier for you to appreciate it.

Summary

Remember that probability gives us a measure as to what the 'chance' or 'likelihood' of an event happening is. If we roll a dice, we can expect that one in every six times, we'll roll a six.

This *doesn't* mean that if you roll a dice six times, you are guaranteed to roll a six. It means that in the long run, on average, you'll get a six once every six rolls. Sometimes you might go on a

great run and get three sixes in a row – but other times it might seem like an eternity before you throw a six. Remember, it's all about what you 'expect' to happen.

Luck is a part of life. But if we can estimate our chances, then that at least gives us some sort of idea as to which choice to make, whether a gamble is worth taking, or so forth. When we do have to gamble and take a choice as to what to do, surely it's better to make an informed choice that's more likely to succeed, rather than a random one?

Exercises

1 A normal die (taking values 1 to 6) is rolled. What is the probability that you roll the following? Leave your answers as fractions in their lowest terms where appropriate.

Example: A number greater than 4.

Solution: Of the six possibilities (1, 2, 3, 4, 5, 6) then only two of these (5, 6) are greater than 4, so the probability is $\frac{2}{6} = \frac{1}{3}$.

(a) 4

(b) 1 or 6

(c) An even number

(d) An integer less than 10

(e) Not 2

(f) A negative number

(g) Either 2 or an odd number

(h) A number divisible by either 2 or 3

(i) Either a number less than 3 or a number greater than 3

2 A random letter (from the English alphabet A–Z, with 26 letters) is chosen. A random number from 1 to 5 is also chosen. What is the probability that the following events happen? Leave your answers as fractions where appropriate.

Example: You choose A and 1.

Solution: The probability of choosing A is $\frac{1}{26}$ and the probability of choosing 1 is $\frac{1}{5}$, so multiplying them together the probability is $\frac{1}{26} \times \frac{1}{5} = \frac{1}{130}$

(a) You choose **P** and 2

(b) You choose a letter from A to D, and an even number

(c) You choose a vowel (A, E, I, O, U) and an odd number

(d) You choose any letter apart from Q, and any number apart from 1

(e) You choose any letter and the number 3

3 With three coins, there are eight possibilities – using H for head and T for tail, the eight possibilities are (H, H, H), (H, H, T), (H, T, H), (H, T, T), (T, H, H), (T, H, T), (T, T, H) and (T, T, T). What is the probability of the following?

Example: **If you toss three coins, what is the probability they all come up as heads?**

Solution: **Only one of the eight possibilities is all three heads, so the probability is $\frac{1}{8}$.**

(a) If you toss three coins, what is the probability of getting three tails?

(b) If you toss three coins, what is the probability of getting two heads and one tail?

4 *Extra question – you don't have to do this, but please try it if you are interested!*

Write down all the possibilities for tossing four coins (try to be systematic!). I will tell you that there are 16 possibilities. Having written down all the possibilities, answer the following:

(a) If you toss four coins, what is the probability of getting four heads?

(b) If you toss four coins, what is the probability of getting two heads and two tails?

OTHER TOPICS

17 Areas and volumes

Calculating the areas and volumes of basic shapes

It's often necessary to work out the area or volume of a shape. For example, if you want to build a wall, tile a path, or fill a pond with water, then you need to work out how many bricks/tiles/litres of water to use (probably best to get those in the right order and not fill your pond with bricks!). This is a really common sort of problem, so in this chapter we will work through formulae that will give us the area or volume of some of the more common shapes you might encounter. There are some more complex formulae for some of these but what's given below is all you need to know for now.

Key topics

→ Basic shapes
→ Areas
→ Volumes

Key terms

shape area volume rectangle square parallelogram triangle circle semicircle pi cuboid cube cylinder sphere

smart tip

A topic like this appears mundane but it's all around you. Look around now – you probably see walls. How did the builders (or you!) know how many bricks to buy, how much paint to buy, and so on? They have to use area to calculate what they need, so they don't buy too little or too much (costing money). Remember this – maths has been used to create pretty much everything around you!

Units of measurement

In this chapter I am not going to specify the lengths etc. in terms of what they are being measured out of. If the numbers represent lengths in metres (m) then the areas will be in square metres (m^2) and the volumes will be in cubic metres (m^3) – similarly if the numbers represent centimetres (cm) then the areas will be in square centimetres (cm^2) and the volumes will be in cubic centimetres (cm^3). Because you could measure out of anything, I'm just going to do everything with simple numbers – you can imagine for yourself measuring it in metres, or centimetres, or inches, or whatever, if you want to imagine it practically. The appendix ('some useful things') gives you some measurement units you might come across.

→ Areas

Area is just a measurement of what is 'inside' a shape.

Rectangles

A *rectangle* is any shape with four 'right angles' in it (a right angle is 90°, or when two lines are 'perpendicular' – they make an exact 'corner' with no slope) like the figure below, which represents a rectangle of *width* (how far across) 8 and *height* (how far up) 4.

You often use the word 'by' to describe a rectangle – this rectangle is said to be '8 by 4' which is an abbreviation for a rectangle of width 8 and height 4.

Suppose this represented an area of ground (8m by 4m) and you wanted to put in enough 1m square tiles to tile it completely. You need eight tiles to go across, and there are four rows, so in total you need $8 \times 4 = 32$ tiles. This gives us the general idea of the area of a rectangle – we just multiply the width by the height to get our answer.

So the area of the above rectangle is $8 \times 4 = 32$. Similarly, the area of a rectangle of width 10 and height 3 is $10 \times 3 = 30$, and so on.

The area of a rectangle is its width multiplied by its height.

Squares

Squares are easy as they are just a special type of rectangle, where the width and height are equal.

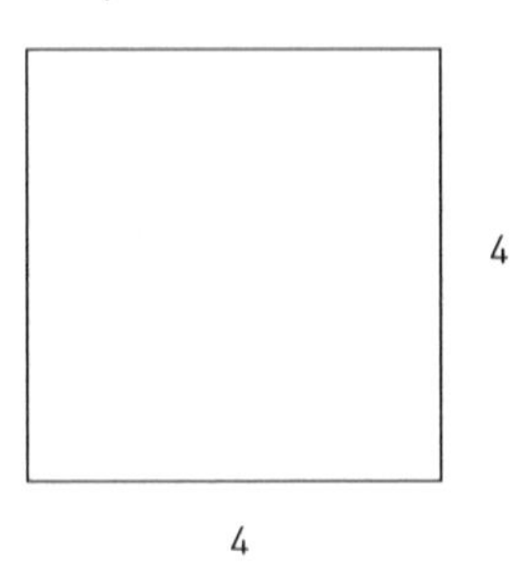

In the above picture, the width and height are both 4, and hence the area is $4 \times 4 = 16$.

Note that you would often write 4×4 as 4^2 (4 squared) – so for squares:

The area of a square is the width squared (multiplied by itself).

We are using 'square' in two meanings here – one for the shape, and one for the act of putting things to the power 2 – don't get confused!

People often refer to 'sides' of a shape – instead of saying 'width 4' you might say 'side length 4'.

So, for example, the area of a square of side length 7 is 49, since $7 \times 7 = 49$ (or you might write $7^2 = 49$).

Parallelograms

A *parallelogram* (don't worry too much about the word, it means that opposite sides are what's called *parallel*, they go in the same direction) is a sort of 'bent rectangle' which looks like below:

What is the area of this? Whilst I will give you the formula and that's all you need to know, it's nice to know where it comes from.

Suppose I cut off the end of my parallelogram as follows:

and moved it to the beginning like this:

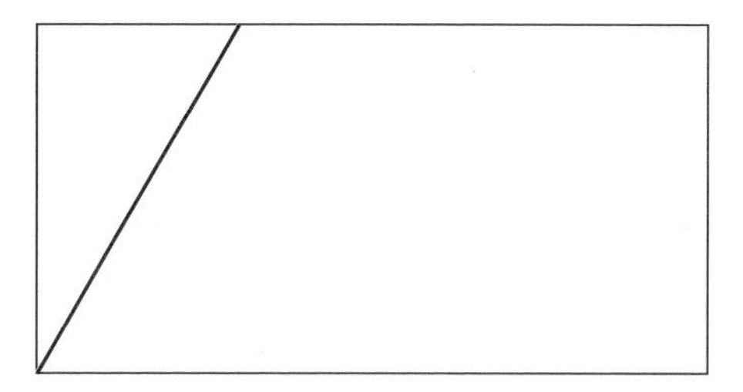

then I've just got a rectangle left with the same area, and you know the area of this is the width multiplied by the height.

Hence the area of a parallelogram is the width multiplied by the 'height' where we take the height here to be the vertical height illustrated below.

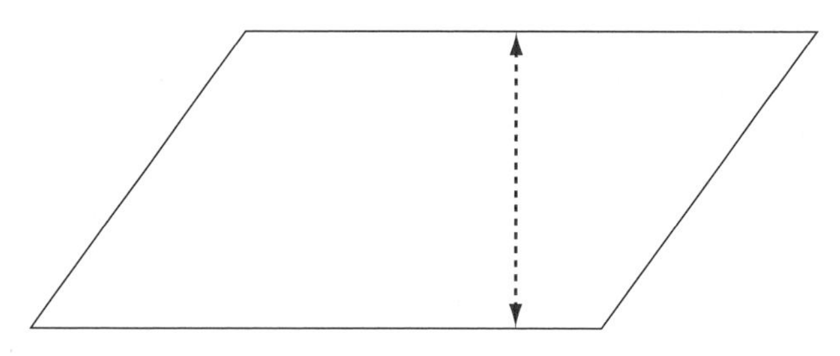

The area of a parallelogram is the width multiplied by the vertical height.

For example, the area of a parallelogram with width 6 and vertical height 4 is $6 \times 4 = 24$.

For your information (in case you ever come across it) a *rhombus* is a special type of parallelogram where the sides are all of the same length (like a 'bent square') – so you can just work out its area in the same way (width multiplied by vertical height).

Triangles

A *triangle* is a shape with three sides. We'll start by looking at right-angled triangles, where one of the angles is a right angle (as discussed before) like the figure below.

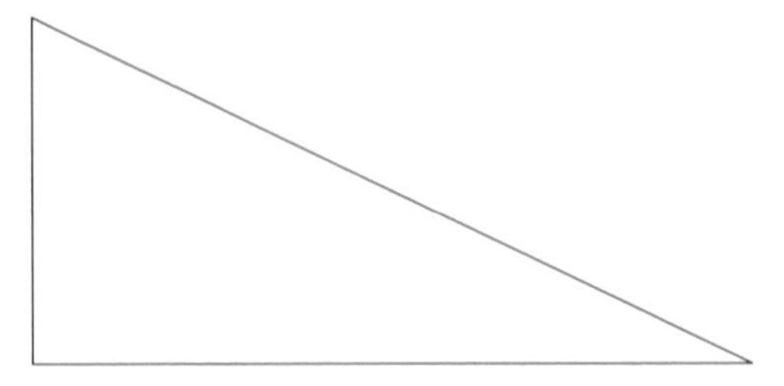

Notice that if you take a perfect copy of this triangle, turn it round, and put it on top you get a rectangle.

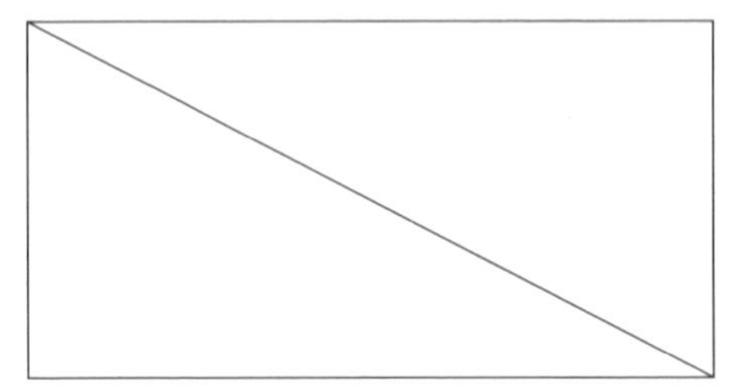

So the area of two right-angled triangles is just worked out by multiplying the width by the height, as in a rectangle.

Therefore (since two triangles make this answer), the area of the original one triangle is half of this.

Hence:

The area of a right-angled triangle is half of the width multiplied by the height.

So for example the area of a triangle with width 6 and height 3 is $\frac{1}{2} \times 6 \times 3 = 9$.

This formula works as well for general triangles – it's not quite as obvious but if you take the vertical height to be as shown below (like in a parallelogram), you get the same formula for the area.

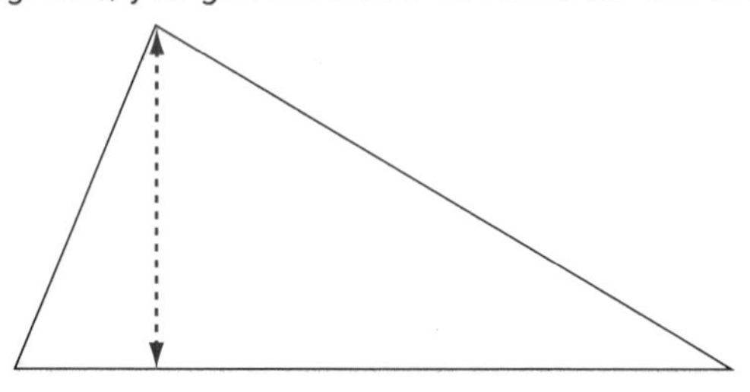

The area of a triangle is half of the width multiplied by the vertical height.

> Some people learn this as '**half the base times the height**' which is exactly the same and it's fine if you have learnt it that way – base just means 'width' here.

So for example, the area of a triangle with width 8 and vertical height 5 is:

$$\frac{1}{2} \times 8 \times 5 = 20$$

Circles

You know what a circle is, but there are a few useful definitions to know about circles.

- The *centre* is the point right in the middle.

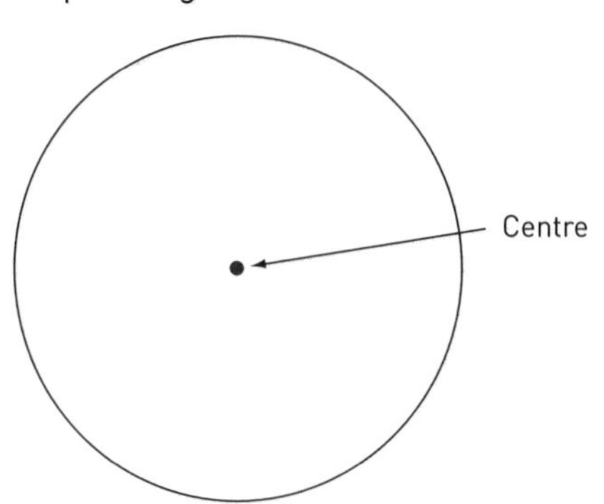

- The radius is the distance from the centre to the circle (*this is the definition of a circle really – it's everything a fixed distance away from a given point*).

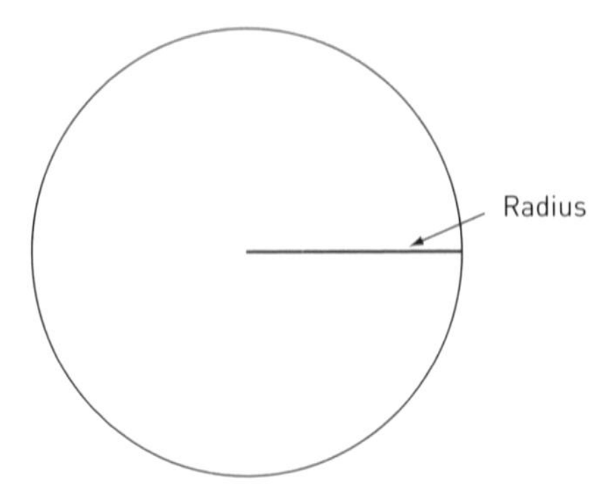

- The *diameter* is the distance across the circle, going through the centre – note that this is twice the radius.

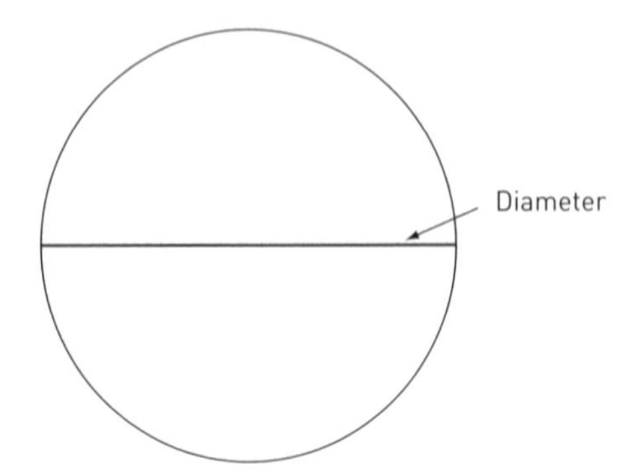

- The *circumference* is the distance all around the circle.

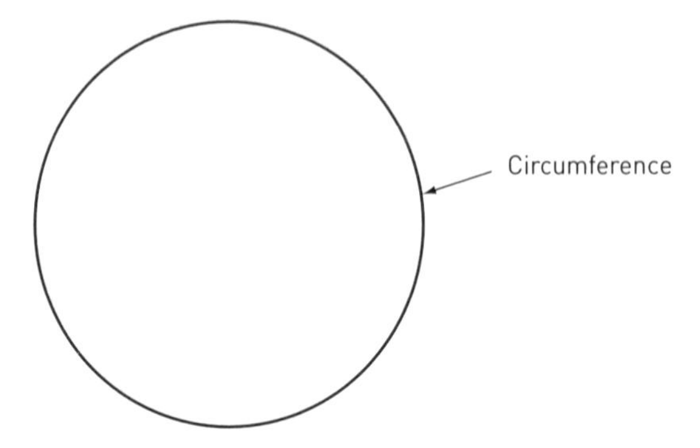

There are some formulae that you should know about circles.

Suppose the radius is denoted by r. Also recall the number π (pi) which we discussed in Chapter 5. It's a really important number in lots of branches of maths, but especially useful for you here with circles. You can take it to be approximately 3.142, though your calculator can probably give it more accurately if it has a π button. Then:

The circumference of the circle is given by $2\pi r$

The area of a circle is given by πr^2.

For example, the circumference of a circle of radius 6 is $2\pi \times 6$, which will give about 37.699, and the area of the same circle is $\pi \times 6^2$, which is around 113 (the exact answer you get on your calculator will depend on how accurate a value of π you use).

Semicircles

A semicircle is just half a circle:

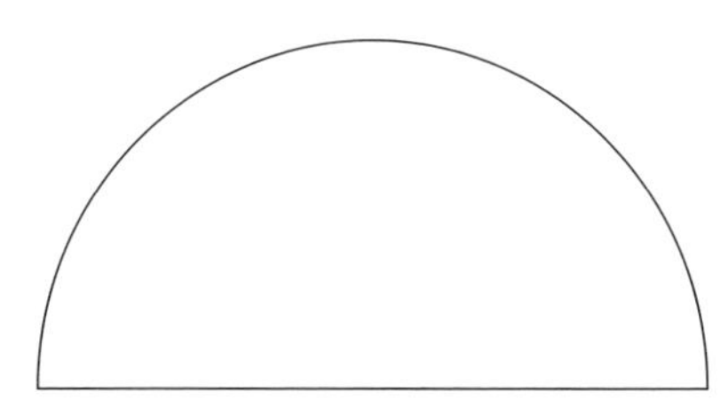

Since it is just half a circle, it's immediate that its area is half that of a circle:

The area of a semicircle is given by $\frac{1}{2}\pi r^2$.

So, for example, the area of a semicircle of radius 6 is $\frac{1}{2} \times \pi \times 6^2$ which works out to be about 56.5 (again the actual answer you will obtain depends on what you use for π and how accurate it is) – note that this is half of the answer we got for the full circle above, which is what we would expect.

→ Volumes

All of the shapes discussed so far are two-dimensional shapes – that is, 'flat' – and we've measured their area. In three dimensions there are many more shapes – when we measure what is 'inside them' we are now looking at this in three dimensions. This is called *volume*.

Volume is a measure of what is 'inside' a three-dimensional shape.

Cuboids

A cuboid is like a three-dimensional rectangle – like a brick. As there are three dimensions, we have three lengths to refer to – width, height and depth are common terms used.

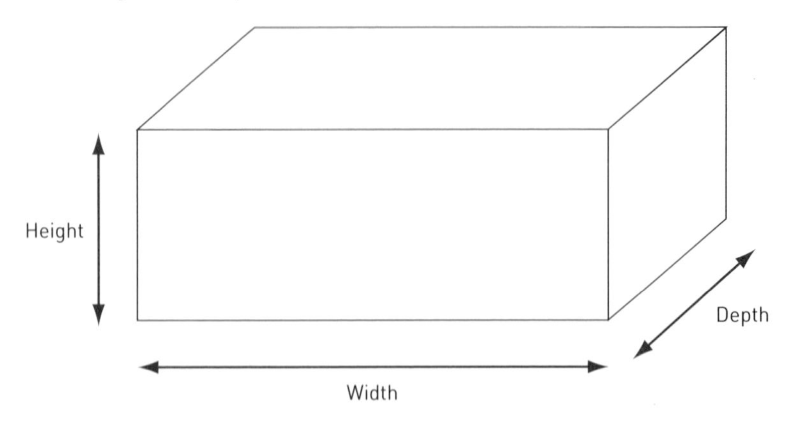

With a rectangle you just multiplied the two side lengths (width and height) together. To work out the volume of a cuboid, it's the same thing, but you multiply all three side lengths (width, height, depth) together.

The volume of a cuboid is the length of the three sides multiplied together (width $\times$ height $\times$ depth).

For example, the volume of a cuboid with side lengths 6, 3 and 2, is $6 \times 3 \times 2 = 36$.

Again, we might refer to this cuboid as a '6 by 3 by 2' cuboid.

Cubes

A *cube* is a cuboid where all the sides are the same length.

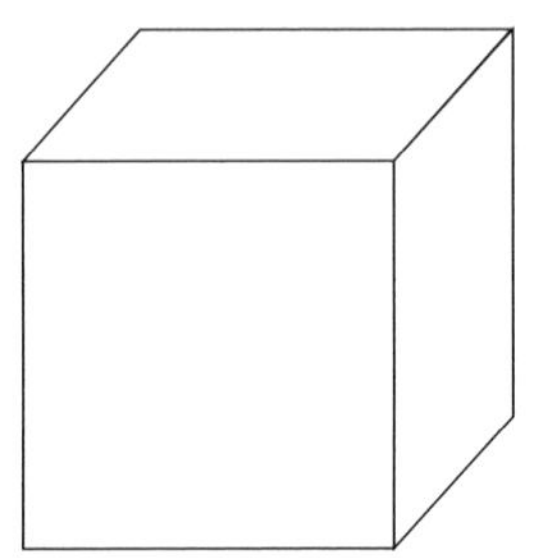

So for a cube, the width, height and depth are all the same, so you just multiply the same thing together three times (called *cubing* the number).

The volume of a cube of side length x is x^3.

For example, the volume of a cube of side length 2 is $2^3 = 2 \times 2 \times 2 = 8$.

Cylinders

A *cylinder* is a shape like a tin of beans – with a circle at each end. Suppose h is the height of this cylinder, and r is the radius of the circle, as shown below.

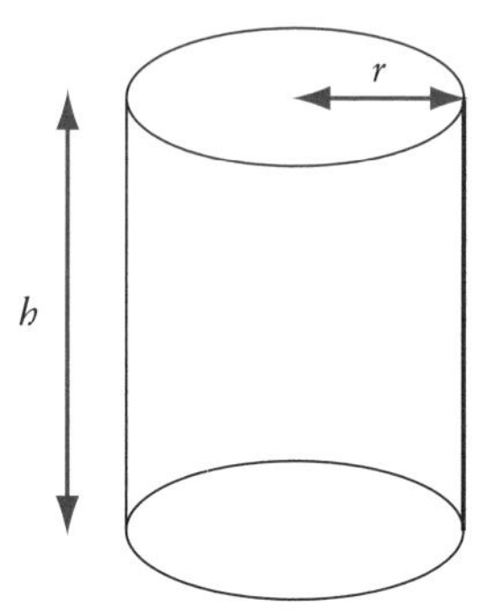

Then since the area of a circle is πr^2, then it makes sense that the volume of the cylinder is obtained by multiplying this area by h.

The volume of a cylinder is $\pi r^2 h$

where r is the radius of the circle and h is the height.

Spheres

A *sphere* is a shape like a ball – a three-dimensional circle if you like. This is really hard to draw on a page, so just pick up a ball or something – it has a radius, just like a circle does, from the centre to the edge. Whilst I won't even try to prove this or explain why, this is the formula for the volume of a sphere:

The volume of a sphere is $\frac{4}{3}\pi r^3$

where r is the radius.

So, for example, the volume of a sphere of radius 6 is $\frac{4}{3} \times \pi \times 6^3$ which, on a calculator, works out to be approximately 904.8 (again the accuracy depends on what you used for π).

Other shapes

There are lots of other shapes which have formulae to work out their areas and volumes, but it's impossible for me to give a full list – we have to stop somewhere! If you need the formula for another shape then it's not too hard to find it in a more advanced textbook or on the internet – for example, the volume of a cone (a shape like an ice-cream cornet) is $\frac{1}{3}\pi r^2 h$ where r is the radius of the big circle and h is the height. But I wouldn't expect you to learn all of these – it's just helpful to know where to look!

smart tip

There are lots of formulae here and it's not really realistic to learn them all. Some (like for a rectangle) are quite straightforward but I wouldn't expect that if (in the highly unlikely event it happened!) someone stopped you on the street and asked you for the volume of a sphere, that you'd be able to give the right answer. But you could make a list of all the formulae you need and keep it somewhere where you know where to access it, or know which book or website to look at, so that you can find it if you do need it. This is true for many of the harder to learn things in maths – if you can't remember them, know where to look!

Summary

Areas and volumes are really important and are all around you. Every single wall, lick of paint, laying of stones, has involved calculating how much is needed.

It's really astonishing in a way – look around you and see how much maths has contributed to the environment – every object has an area or volume, and someone needs to know what that is – so this topic is literally everywhere!

Don't stress about learning all the formulae – it's more important that you understand them and know where to look if you do need them.

Exercises

Use the π button on your calculator if you have one, but otherwise just use 3.142.

1 Calculate the area of the following shapes.

Example: A triangle of width 8 and vertical height 3.

Solution: The area of a triangle is given by $\frac{1}{2}$ multiplied by the width multiplied by the height, so the answer is $\frac{1}{2} \times 8 \times 3 = 12$.

(a) A rectangle of width 9 and height 5

(b) A square of side length 8

(c) A parallelogram of width 7 and vertical height 5

(d) A triangle of width 5 and vertical height 3

(e) A circle of radius 5

(f) A semicircle of radius 2.5

(g) A circle of diameter 12 (work out the radius first)

2 Calculate the volume of the following shapes.

Example: A cuboid with width 5, height 3 and depth 4

Solution: The volume of a cuboid is obtained by multiplying the width, height and depth together so you get $5 \times 3 \times 4 = 60$.

(a) A cuboid with width 7, height 3, and depth 4

(b) A cube with side length 5

(c) A cylinder with height 8 and radius 3

(d) A sphere of radius 3.5

(e) A cone of height 10 and radius 2

3 Answer the following questions.

Example: How many 0.5m by 0.5m tiles are needed to tile a 6m by 4m wall?

Solution: You need 12 tiles to go across ($12 \times 0.5 = 6$) and 8 tiles to go up ($8 \times 0.5 = 4$) and so in total you need $12 \times 8 = 96$ tiles.

(a) How many 1cm by 1cm tiles do you need to tile a 5cm by 5cm grid?

(b) How many 0.5m by 0.5m paving stones are needed to pave a patch of ground 3m by 2m?

(c) How many 0.5cm by 0.5cm by 0.5cm cubes can fit in a cuboid which is 2cm by 3cm by 4cm?

18 Logarithms

Calculating logarithms - the opposite of powers

For some reason, the very word 'logarithm' sounds hard. Maybe it's the way it's spelt, or the 'old-fashioned' feeling it conjures up, but it just seems difficult.

Actually logarithms are very closely related to powers and indices, which you already studied, and aren't really anything to be scared of.

I strongly recommend you revise Chapter 7 before doing this chapter!

Key topics

→ Logarithms
→ Laws of indices
→ Laws of logarithms
→ Using your calculator

Key terms
logarithm log base laws of indices laws of logarithms natural logarithm calculator.

→ Definition of a logarithm

Suppose I asked you to work out what 2^3 is. Recalling Chapter 7, you should know that 2^3 is the same as $2 \times 2 \times 2 = 8$.

Let me put this question in another way. Rather than saying 'what is 2^3' and getting the answer 8, I'll ask the question 'what power do you have to put 2 to, to get the answer 8?' Clearly the answer is 3, since $2^3 = 8$.

The first question is an indices question, the second a logarithms question.

You can informally remember this as follows:

$2^3 = ?$ (this is an indices question with answer 8)

$2^? = 8$ (this is a logarithms questions with answer 3)

Note that it's basically the same question, just we're asking for something else – with indices we want the final answer, with logarithms we've got the answer and we want to know the power.

This can be formalised as follows, which also introduces the notation we use for logarithms. We use the abbreviation *log* to stand for logarithm.

The answer to $\log_a(b)$ is whatever power you need to put a to, to get the answer b.

Here a is called the *base* of the logarithm.

So, for example, to work out $\log_2(8)$, you need to answer the question $2^? = 8$ (what power do I have to put 2 to, to get the answer 8?), and the answer is 3, so you would write $\log_2(8) = 3$.

Some other examples:

- $\log_{10}(100) = 2$ since $10^2 = 100$
- $\log_2(16) = 4$ since $2^4 = 16$
- $\log_{10}(10000) = 4$ since $10^4 = 10000$
- $\log_3(27) = 3$ since $3^3 = 27$

smart tip

Understand how different branches of maths come together. You probably never thought before that logarithms had anything to do with powers, but they're almost the same – just asking the same question in the 'opposite' way. It's not meant to be hundreds of different subjects – everything comes together and if you can start to understand how one topic relates to another you get a much greater understanding of maths as a whole and get much more confidence!

→ Logarithms – further examples

The above examples are most of what you will need, but you should see some further examples.

For example, what is $\log_{10}(1)$? This is harder to think about. But remember that this is asking the question $10^? = 1$, and then remember the rule in Chapter 7 that anything to the power 0 is 1. Hence, the power we are looking for is 0, since $10^0 = 1$, so we can say that $\log_{10}(1) = 0$.

Also, what is $\log_2(\frac{1}{8})$? Again, this is quite hard. But remember about negative powers – you might recall that $2^{-3} = \frac{1}{2^3} = \frac{1}{8}$. So the power we are looking for is -3.

Hence $\log_2(\frac{1}{8}) = -3$, since $2^{-3} = \frac{1}{8}$.

As a final example here, what is $\log_{16}(4)$? Again, this is hard to see directly. Notice though that 4 is the square root of 16, and square roots were represented by the power $\frac{1}{2}$. Hence, since $16^{1/2} = 4$, the power we are looking for is $\frac{1}{2}$ and so $\log_{16}(4) = \frac{1}{2}$.

A summary of these is below:

- $\log_{10}(1) = 0$ since $10^0 = 1$
- $\log_2(\frac{1}{8}) = -3$ since $2^{-3} = \frac{1}{8}$
- $\log_{16}(4) = ½$ since $16^{1/2} = 4$

These ones are harder to do and really you just need to have a good intuition and knowledge as to what power to think of – there's no real 'way' to do it, you just have to think. As advised before – practise, practise, practise! You can find more examples in the exercises, on the website, on other sources, or just make your own – but the more you practise the better you get!

smart tip

Work together and help and challenge each other! It can often be a lot easier to master a topic if you sit down with a friend and work through it – you can help each other with the bits one of you finds hard, but the other understands. For example with logarithms, try creating your own example and testing your friend – you'll learn by creating the example and you'll learn from your friend's response, and they'll learn by doing it. When learning a topic, working together can be really beneficial as you can feed off each other and make much faster progress together.

→ Laws of indices and logarithms

Remember that with indices, we had rules like $2^3 \times 2^4 = 2^7$ – where we got the answer 7 by adding together the two indices 3 and 4. These rules can be written algebraically to show that they work for any numbers:

- $x^a x^b = x^{a+b}$ (note that we don't write down the multiplication sign $\times$ in algebraic expressions)
- For example, $2^5 \times 2^7 = 2^{12}$ and $5^3 \times 5^{-7} = 5^{-4}$
- $\frac{x^a}{x^b} = x^{a-b}$.
- For example, $\frac{2^7}{2^3} = 2^4$ and $\frac{5^4}{5^{-5}} = 5^9$
- $(x^a)^b = x^{ab}$
- For example, $(2^3)^4 = 2^{12}$ and $(2^4)^{-2} = 2^{-8}$

There are similar laws for logarithms. I'll give you these algebraically and then we'll illustrate with examples.

- $\log_a(xy) = \log_a(x) + \log_a(y)$
- $\log_a(\frac{x}{y}) = \log_a(x) - \log_a(y)$
- $\log_a(x^y) = y \times \log_a(x)$

Examples of each of these are as follows:

- $\log_2(4 \times 8) = \log_2(4) + \log_2(8)$

You can check this is right. 4×8 is 32, so this says that $\log_2(32) = \log_2(4) + \log_2(8)$.

But $\log_2(32) = 5$ (since $2^5 = 32$), $\log_2(4) = 2$ (since $2^2 = 4$), and $\log_2(8) = 3$ (since $2^3 = 8$). So this says that $5 = 2 + 3$ which is correct, and we checked that it seems right at least for this example – in fact it always works.

Check for yourself the following:

- $\log_2(\frac{32}{8}) = \log_2(32) - \log_2(8)$
- $\log_2(4^3) = 3 \times \log_2(4)$

These laws will always work. It's much easier to learn them by idea (as suggested in the tip below) than try to learn all the symbols.

It's hard (almost impossible?) to learn these rules symbol by symbol, but you should learn them by what their idea is. It's much easier to learn the concept ('if you multiply powers then you add the indices together') than it is to learn the symbols, and it makes more natural sense. When you have to learn formulae, then try to learn them in this way rather than trying to learn every symbol.

→ Logarithms on your calculator

I obviously don't know what calculator you have, but most scientific calculators have buttons built in to them to do logarithms.

You may or may not have a button that can do general logarithms (don't worry if not) – but if you have a scientific calculator, you probably do have two buttons labelled *log* and *ln* (that's short for 'logarithm natural' so l followed by n – if you thought that was actually *in*, don't worry, it's a common misunderstanding!)

The *log* button will work out the logarithm to base 10 – this is a very common base to use. So for example if you use the *log* button with 1000 you should get the answer 3, since $\log_{10}1000 = 3$ (since $10^3 = 1000$).

The *ln* button works out something called the *natural logarithm* which is a logarithm to base *e*. What is *e*? This is one of the 'special' numbers in maths (a bit like π) – it's approximately 2.718281828 but like π, it goes on forever, with no predictable pattern. This number is hugely important in more advanced maths (like calculus) which is why your calculator has it built in – don't worry about it for now but it's nice to know what it does!

Advanced calculators have buttons to do general logarithms to any base – if you do end up needing this in your studies then of course you should invest in a more powerful calculator, but don't worry about it if you don't need it – stick to what you need!

smart
tip

Get to know your calculator! If you don't know what a button does, find out. Probably it does something you don't need, but at least you know that now. If it actually does something useful then you know how to do it from now on – either way you haven't lost anything. The same tip applies for anything really – if you understand what everything does, you can use the relevant things and not worry about the unimportant things (and at least know why they are not important).

Summary

A topic like this really shows the connection between different branches of maths. Really, this is no different from the indices chapter – just instead of asking what the answer is to a question like x^y, we are asking what power we need to get a particular answer.

You should keep this in mind, and remember that every subject isn't distinct – the topics do come together and are used together.

Indices are useful (you often need to do repeated calculations) – so if indices are useful, so are logarithms, since it's all about the same thing!

Exercises

Try to do question 1 without a calculator!

1 Calculate the following logarithms.

Example: $\log_2(16)$

Solution: Since $2^4 = 16$, the answer is 4 (the index), so $\log_2(16) = 4$.

(a) $\log_2(8)$ (b) $\log_3(9)$ (c) $\log_4(16)$ (d) $\log_2(32)$

(e) $\log_5(25)$ (f) $\log_2(64)$ (g) $\log_{10}(1000)$ (h) $\log_2(1)$

(i) $\log_3(\frac{1}{3})$ (j) $\log_2(\frac{1}{16})$ (k) $\log_9(3)$ (l) $\log_{49}(7)$

2 Check that the following are true:

Example: $\log_2(16 \times 4) = \log_2(16) + \log_2(4)$

Solution: $16 \times 4 = 64$, and $\log_2(64) = 6$ since $2^6 = 64$. $\log_2(16) = 4$ since $2^4 = 16$, and $\log_2(4) = 2$ since $2^2 = 4$, so we have $6 = 4 + 2$ which is correct.

(a) $\log_2(8 \times 8) = \log_2(8) + \log_2(8)$

(b) $\log_3(\frac{27}{9}) = \log_3(27) - \log_3(9)$

(c) $\log_{10}(100^2) = 2 \times \log_{10}(100)$

3 *Extra question*:

Can you take the logarithm of a negative number, for example $\log_2(-8)$? If not, why not?

19 Quadratic equations

Solving equations involving powers of x

A *quadratic equation* is any equation that has x^2 in it (the word quadratic refers to power of 2). These are much more difficult to solve as none of the techniques we used before help – how on earth do you come up with a solution for something like $x^2 + 5x + 6 = 0$? That's what we will investigate here.

Key topics

→ Quadratic equations
→ Factorisation
→ The quadratic formula

Key terms

quadratic equation factorisation quadratic formula
positive and negative square roots polynomial

→ Brackets and factorisation

Recall what we did before when we multiplied out brackets. Try to work out

$$(x + 2)(x + 3)$$

What do you get?

If you do it right, remembering how to multiply out the brackets and then collect together like terms, you should get $x^2 + 5x + 6$

If you can't get this – go back and revise Chapter 9 (Brackets in algebra) before you return to this chapter.

So we have $x^2 + 5x + 6 = (x + 2)(x + 3)$. Notice a couple of things here:

(i) If you multiply together the two numbers in the brackets, you get $2 \times 3 = 6$, which is the number at the end of our 'quadratic' expression $x^2 + 5x + 6$.

(ii) If you add together the two numbers in the brackets, you get $2 + 3 = 5$, which is the number in the middle of our 'quadratic' expression $x^2 + 5x + 6$.

Our aim is going to be to take an expression like $x^2 + 5x + 6$ and convert it into the form $(x + 2)(x + 3)$. Going backwards like this is called *factorisation*.

This is going to help us solve such equations. We will follow the technique we have suggested here to take an expression with x^2 in it and 'factorise it back' into an expression with two brackets. You'll see the reason for us to do this shortly, but for now let's just do it!

So, following what we did above, given an expression like $x^2 + 5x + 6$ you have to think of two numbers that:

- multiply together to make the last number (6)
- add together to make the middle number (5)

To be honest, there's no quicker way of doing this than just thinking about it and trying to find the numbers.

After thinking, you can come up with the numbers 2 and 3 – they multiply together to make 6 and add together to make 5. So this is going to be the same as $(x + 2)(x + 3)$.

Let's look at another few examples to make this clear.

Question: Factorise $x^2 + 6x + 8$

Solution: Think of two numbers that multiply together to make 8, and add together to make 6. After some thought, you should work out that 2 and 4 are the numbers we are after.

Hence this is the same as $(x + 2)(x + 4)$ – check this!

Bear in mind that the numbers could be 1:

Question: Factorise $x^2 + 9x + 8$

Solution: You need to think of two numbers that multiply together to make 8, and add together to make 9. These numbers are 1 and 8.

Hence this is the same as $(x + 1)(x + 8)$ – check this!

The numbers could also be negative – if you have a negative number at the end then one will be positive and one will be negative:

Question: Factorise $x^2 + 3x - 10$

Solution: You need to think of two numbers that multiply together to make -10, and add together to make 3. These numbers are 5 and -2.

Hence this is the same as $(x + 5)(x - 2)$ – check this!

If the last number is positive but the middle number is negative then you will have two negative numbers:

Question: Factorise $x^2 - 5x + 6$

Solution: You need to think of two numbers that multiply together to make 6, and add together to make -5. These numbers are -2 and -3.

Hence this is the same as $(x - 2)(x - 3)$ – check this!

smart tip

Something like this comes with practice. It's not easy to think of the two numbers at first, but the more you practise the better you get and the more intuition you develop so that you can think of the two numbers more quickly – so keep practising!

→ Using factorisation to solve quadratic equations

You can use this technique to solve quadratic equations. For example, what is a solution to $x^2 + 5x + 6 = 0$?

Well, as we did above, note that $x^2 + 5x + 6$ is the same as $(x + 2)(x + 3)$ and so the equation is $(x + 2)(x + 3) = 0$

Why does this help? Well, if two things multiply together to make zero, then one of the two things must be zero. This means that either the first term $(x + 2)$ is zero, or the second term $(x + 3)$ is zero. Hence, either $x + 2 = 0$ or $x + 3 = 0$, from which it follows that either $x = -2$ or $x = -3$.

Check these answers in the original equation – both (-2) and (-3) do work (remember that the square of a negative number will be positive, since 'negative times negative is positive'. For example, $(-2)^2 + 5(-2) + 6 = 4 - 10 + 6 = 0$ – check -3 for yourself.

So there are two answers, both of which work (-2 and -3 here). For quadratic equations in general, there might not be any solutions at all, there might just be one, but normally there will be two different solutions:

A quadratic equation normally has two different solutions.

It might be easier to get your solution via the following technique. Note that the 'two numbers we were looking for' to factorise were 2 and 3, and the answers were -2 and -3 – their 'opposites'. You can always do this:

> ***Example***: Solve the quadratic equation $x^2 + 4x - 21 = 0$
>
> ***Solution***: Think of two numbers that multiply together to make -21 and add together to make 4. The numbers you need are 7 and -3.

Either then write $(x + 7)(x - 3) = 0$ and deduce that either $x + 7 = 0$ or $x - 3 = 0$, and hence $x = -7$ or $x = 3$ . . .

. . . or just take the 'opposites' of the two numbers you thought of (7 and -3) to get the answers of -7 and 3.

Check for yourself that -7 and 3 both work and both give solutions to the equation above.

An example of an equation where there is only one solution is $x^2 + 6x + 9 = 0$. We want to factorise this – so we need two numbers that multiply together to make 9 and add together to make 6. These numbers are 3 and 3, so we have -3 and -3 as our answers – but this is just the same thing, so the only answer is $x = -3$.

If you had $x^2 + 2x + 9 = 0$ there won't be any solutions at all – you can't think of two numbers which multiply together to make 9 and add together to make 2. So an equation like this isn't solvable at all. Fortunately you won't encounter these very often in your studies!

→ The quadratic formula

It's always nice, where possible, to solve quadratic equations using factorisation like this – you get the answer nice and easily. Unfortunately, some equations don't solve so readily. An equation like $x^2 + 9x + 4$ does have solutions, but you won't be able to think of two numbers that multiply together to make 4 and add together to make 9.

There is actually a formula that can help you solve equations like this – you'll need a calculator to do this, and also a sharp intake of breath before I give you the formula.

Suppose you had an equation $ax^2 + bx + c = 0$, where a, b, c are some numbers. Then the solutions to this equation are given by:

$$x = \frac{-b \pm \sqrt{b^2 - 4ac}}{2a}$$

Wait – what on earth is that? OK, I know, it's horrible. Let's do an example to try to illustrate it. Get your calculator ready!

Take the equation above, $x^2 + 9x + 4$.

- a is the number of x^2 that we have – in this case just 1 (remember that x^2 is short for $1x^2$)
- b is the number in the middle, the number of x that we have – in this case 9
- c is the number at the end – in this case 4

So $a = 1$, $b = 9$ and $c = 4$.

Now we're going to put these numbers into the formula. But we've still got a problem – what does that weird symbol $\pm$ mean? This is a plus above a minus – it means we are going to do the calculation once with a plus (+) and once with a minus (−), which means we will get two different answers – remember that we do expect two different answers for most quadratic equations!

The symbol $\pm$ is short for 'plus or minus' and means do the calculation twice – once with plus and once with minus.

So, let's do it with + (plus) first. We need to evaluate $x = \frac{-b + \sqrt{b^2 - 4ac}}{2a}$ with our values – when $a = 1$, $b = 9$ and $c = 4$. Evaluating, you should get $\frac{-9 + \sqrt{9^2 - 4 \times 1 \times 4}}{2 \times 1} = \frac{-9 + \sqrt{81 - 16}}{2}$ $= \frac{-9 + \sqrt{65}}{2}$

At this point you'll probably have to turn to a calculator – you should get −0.469 to 3 decimal places.

If you repeat the calculation again with − (minus) instead, you get

$$\frac{-9 - \sqrt{9^2 - 4 \times 1 \times 4}}{2 \times 1} = \frac{-9 - \sqrt{81 - 16}}{2} = \frac{-9 - \sqrt{65}}{2}$$

which is −8.531 to 3 decimal places.

So our two answers are (approximately) -0.469 and -8.531

Make sure you can use your calculator to get these answers right. I can't help you with your calculator in this book – I don't know what sort of model you have. There are thousands of different calculators, from the basic to the complex – you need to understand how your own calculator works. If you don't know how to do it, ask someone, or look at the manual!

You don't need to know where this formula comes from! It can be proved using a technique called 'completing the square', which is beyond what you need for now – you just need to be able to use it.

To convince you that it does work, try to solve an equation like $x^2 + 4x - 21 = 0$ which we did before (getting the answers -7 and 3). In this case $a = 1$, $b = 4$ and $c = -21$.

Get ready ...

One solution (doing +) is

$$\frac{-4 + \sqrt{4^2 - 4 \times 1 \times (-21)}}{2 \times 1} = \frac{-4 + \sqrt{16 - 4 \times (-21)}}{2}$$

$$= \frac{-4 + \sqrt{16 + 84}}{2} = \frac{-4 + \sqrt{100}}{2} = \frac{-4 + 10}{2} \text{ (since } \sqrt{100} = 10)$$

$$= \frac{6}{2} = 3$$

and we did get the answer of 3 that we expected. I will leave it to you as an exercise to do the calculation with the minus sign instead to get the other answer -7.

Note that this formula can be used when there is a number in front of the x^2. It's very hard to factorise and then solve $2x^2 + 7x + 3 = 0$ but we can use the formula above with $a = 2$, $b = 7$ and $c = 3$ to get our solutions – even though it's quite hard work putting the numbers into the formula and evaluating them, it is doable! Try to do this as an exercise.

→ Polynomials

Just as an aside, equations with powers of x in them are called *polynomials*. You might like to know that there is also a formula to solve equations with powers of x^3 in them, and even with powers of

x^4 in them. But these formulae are horrendous – I can't even really print them here. Search on the internet for 'cubic formula' and 'quartic formula' if you really want to see them – just brace yourself first!

It's interesting, though, that these horrible formulae exist, but there actually isn't a formula for an equation with x^5 in it. It's not that nobody has bothered to do it, it was actually shown that such a formula can't exist at all, so there's no need to bother trying to find it. This was shown in the 1800s by a young French mathematician called Evariste Galois, who managed to leave his mark on mathematics despite dying in a duel at the young age of 20. Galois was a very interesting character who is well worth looking up if you would like to learn more about the history involved in mathematics.

smart tip

Remember that all this maths came from somewhere. It can be interesting to look up some of the names you see, and the maths that you see, to try to understand where it came from. It was real people that created this, through a journey of discovery – it's not just some mad invention, there are real people involved with real stories to tell. Gaining an appreciation of history and the people involved can help to make maths seem real, and not just some obscure thing.

Summary

Equations with x^2 in are hard. But there are techniques available to us to make them possible. This is a lot of what maths is about – applying techniques to solve problems.

Solving these equations by factorisation is much easier if you can actually think of the two numbers needed – but in cases where not, there is a formula that you can fall back on, even if it's not that easy to use, it does exist and will give you the right answers as long as you type it into your calculator correctly.

It is important for you to learn how to use your calculator correctly – it can only do what you tell it, and if you can't tell it the right thing, it can't give you the answer you want.

Exercises

1 Solve the following equations by factorisation.

Example: $x^2 + 9x + 20 = 0$

Solution: You need to think of two numbers that multiply together to make 20 and add together to make 9. These numbers are 4 and 5. So the answers you are looking for are their 'opposites' -4 and -5 (or you can say that $x^2 + 9x + 20 = (x + 4)(x + 5)$ and so either $x + 4 = 0$ or $x + 5 = 0$, so $x = -4$ or $x = -5$).

(a) $x^2 + 7x + 12 = 0$ (b) $x^2 + 8x + 12 = 0$ (c) $x^2 + 6x + 5 = 0$

(d) $x^2 + 4x - 12 = 0$ (e) $x^2 - 2x - 15 = 0$ (f) $x^2 - 6x + 8 = 0$

(g) $x^2 - 8x + 7 = 0$ (h) $x^2 + 8x + 16 = 0$ (i) $x^2 - 2x + 1 = 0$

2 Use the quadratic formula to solve the following equations. If you have to give decimal answers, give them to 3 decimal places.

Example: $2x^2 + 8x + 3 = 0$

Solution: In this case $a = 2$, $b = 8$ and $c = 3$. Remembering to do the formula twice, first we get

$$\frac{-8 + \sqrt{8^2 - 4 \times 2 \times 3}}{2 \times 2} = \frac{-8 + \sqrt{64 - 24}}{4} = \frac{-8 + \sqrt{40}}{4}$$

which is -0.419 to 3 decimal places, and secondly we get

$$\frac{-8 - \sqrt{8^2 - 4 \times 2 \times 3}}{2 \times 2} = \frac{-8 - \sqrt{64 - 24}}{4} = \frac{-8 - \sqrt{40}}{4}$$

which is -3.581 to 3 decimal places. So the two solutions are -0.419 and -3.581.

(a) $3x^2 + 5x + 1 = 0$ (b) $2x^2 - 7x + 3 = 0$ (c) $x^2 + 9x + 2 = 0$

(d) $2x^2 + 5x - 2 = 0$ (e) $3x^2 - x - 1 = 0$ (f) $4x^2 + 4x + 1 = 0$

3 An equation like $x^2 + 3x + 5 = 0$ has no solutions. What goes wrong if you try to use the quadratic formula with this equation?

20 An introduction to trigonometry

Calculations with triangles, including angles and trigonometric functions

This chapter will give you a gentle introduction to the concept of trigonometry – calculations with triangles. I won't go far – if you need something deeper then refer to a more advanced textbook, but I want to give you just some of the basic ideas so you get an idea of what this topic is all about.

The whole branch of maths concerned with shapes is called *geometry* but here we are only interested in triangles, which is *trigonometry*.

Triangles are really important, especially in practical subjects like engineering – again, the concepts may seem abstract to you at a first glance but they are really important!

Key topics

→ Pythagoras' theorem
→ Sine, cosine and tangent

Key terms

triangle right-angled triangle hypotenuse Pythagoras' Theorem angles sides adjacent opposite sine cosine tangent

→ Right-angled triangles

I'm going to focus only on right-angled triangles, where one of the angles is a right angle (90°), so something like the figure below. Note that we often use a square symbol in the angle to denote a right angle.

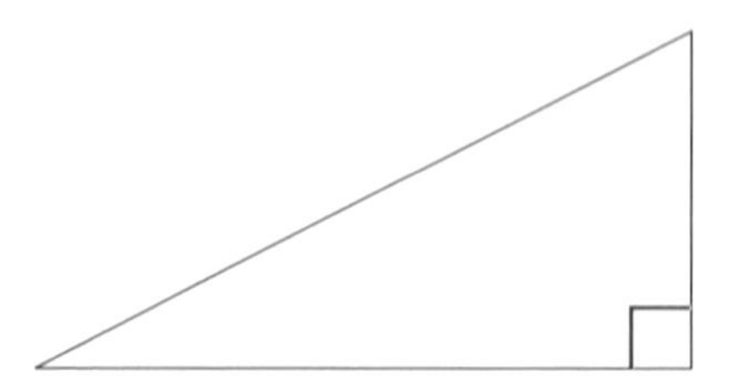

Why focus on these? Well, they are particularly useful. Although in more advanced work you can generalise to other triangles without right angles in them, these crop up very often. For example in the picture above, you could imagine that the vertical line is a wall and the horizontal line is the ground, and the sloping line is a ladder, or a bridge, or something similar.

smart tip

Start to spot shapes around you. If a ladder leans against a wall, note that it makes a right-angled triangle. See if you can see triangles around you (for example, a beam in a roof probably forms a right-angled triangle) and realise that this is really practical stuff – someone used trigonometry to work out the length of beam they needed (for example).

→ Pythagoras' theorem

The following theorem from Pythagoras is one of the best-known theorems in the world. Before I present the theorem and explain why it's useful, a quick definition:

The *hypotenuse* is the longest side in a right-angled triangle.

If you draw the triangle like above, this is the sloping side. It's a word worth knowing because it appears very often in trigonometry – it would be nice if it was an easier word to learn, but it isn't, so you do just have to learn it. By the way, if you don't know how to pronounce it, it's something like 'hi-pot-a-news'. It is worth learning how to say new words (look on the internet or ask someone) so that you don't sound silly if you ever have to say them!

Have a look at the following triangle, where I've labelled the lengths of each side apart from the hypotenuse.

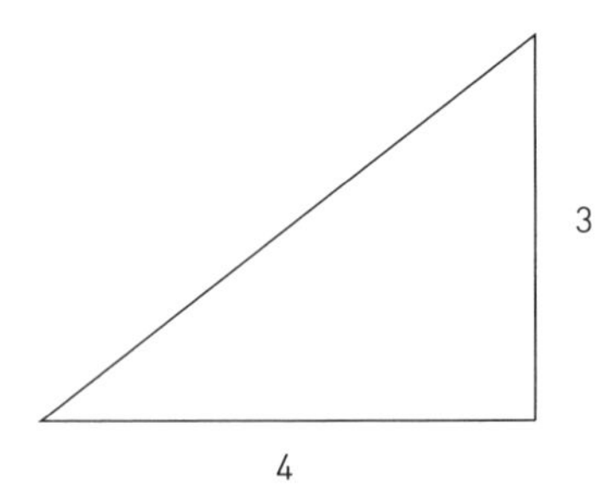

Pythagoras' theorem gives me a way to work out the length of the hypotenuse. This is clearly useful – say for example I wanted to work out the length of the ladder I need to reach the top of a 3m wall from 4m away.

Pythagoras' theorem is as follows, which I will then illustrate with an example:

If you take the square of the two sides and add them together, this is the square of the hypotenuse.

It's perhaps easier expressed in another form – if the other two sides of the triangle are of length a and b, and h is the hypotenuse, then Pythagoras' theorem says:

$$h^2 = a^2 + b^2\text{, or alternatively, } h = \sqrt{a^2 + b^2}$$

In this example, our other two sides are 3 and 4. This means that our hypotenuse is given by

$$h = \sqrt{3^2 + 4^2} = \sqrt{9 + 16} = \sqrt{25} = 5$$

Hence the hypotenuse, the length of the long side, is 5. So in my example above (a 3m wall from 4m away) I need a 5m ladder.

On this occasion, the answer 5 was a nice simple number – unfortunately it won't usually be and you'll have to resort to a calculator.

For example, with the triangle

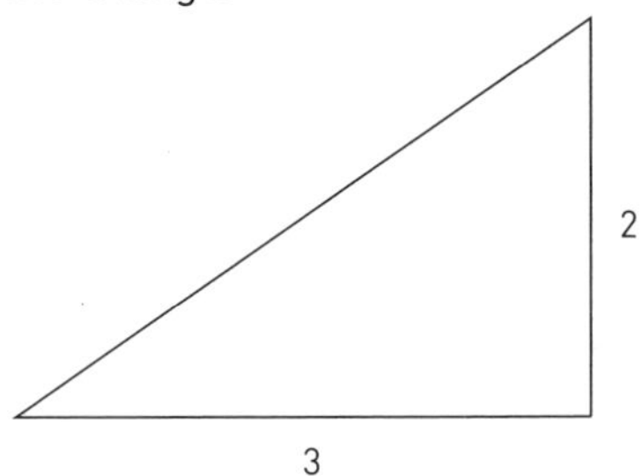

then the hypotenuse is $\sqrt{2^2 + 3^2} = \sqrt{4 + 9} = \sqrt{13}$. Now you'll have to do this on your calculator – you get roughly 3.606.

Note that Pythagoras lived around 2500 years ago, but this formula is still used by thousands of people (especially in construction and design) every day. Even though there is some dispute as to whether he actually created and proved the formula all by himself or whether it just got attributed to him, the name is there and millions of people know his name and this theorem even 2500 years later – that's quite a legacy he left!

→ Angles

First of all a note on angles and sides:

Angles are measured in degrees, so for example 90° (the small superscript circle sign represents degrees). A right angle is 90°.

So a right angle is 90°, as shown below.

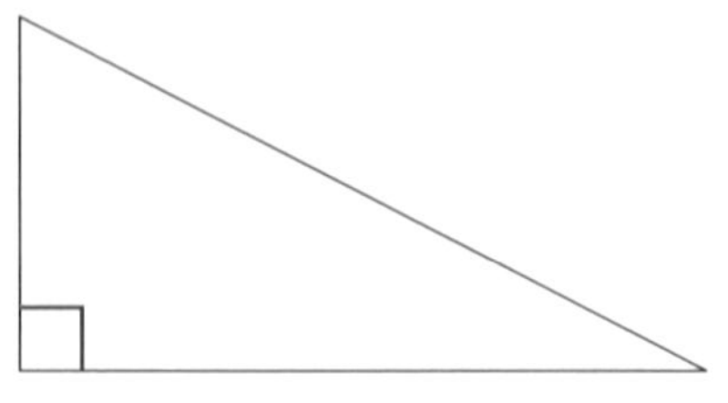

Other angles are often indicated by a curved line rather than the square used for right angles. 45° is 'half of' a right angle and represents an angle shown below.

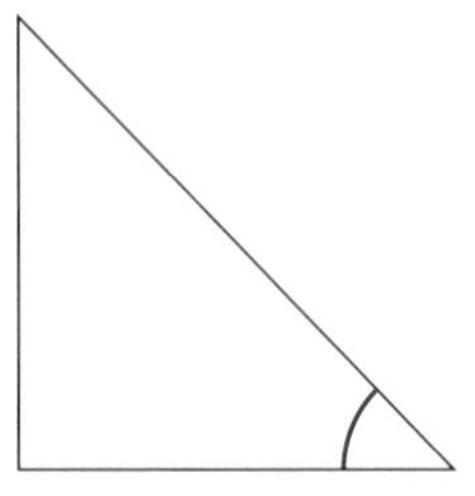

It's possible to have angles bigger than 90° – see for example the following triangle, where the angle is about 135°.

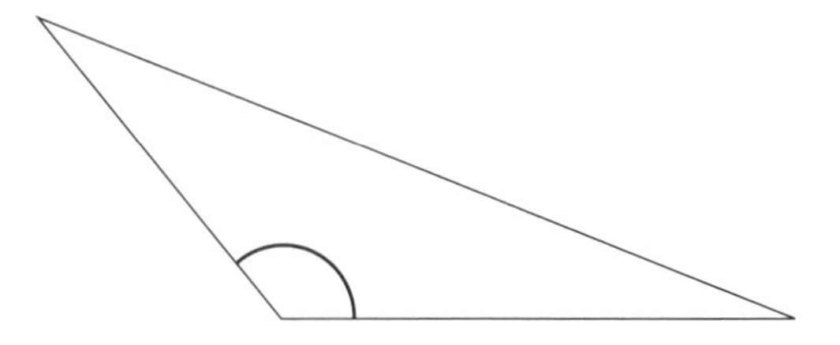

Why do we choose to measure out of 90? No particular reason to be honest, we just have got used to this over time and now people are comfortable with it. There are other ways of measuring angles out of something else (your calculator probably does *radians* where 180° = π radians, which does have advantages when it comes to areas etc., but this is *not* something you need to know about at this stage).

A useful thing to know about angles in a triangle is:

the angles in a triangle always add up to 180°.

For example, the angles in the triangle below are 90°, 60° and 30°, which add up to 180°.

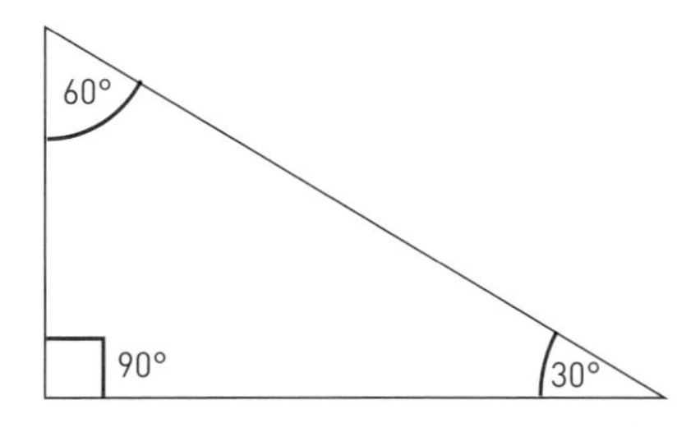

→ Sides

Given a particular angle, you can label the sides as shown in the figure below.

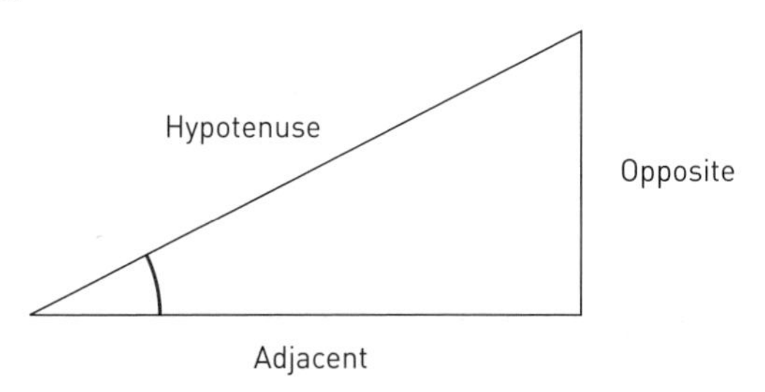

The *hypotenuse* is the longest side, The *opposite* is 'opposite' the angle, and the *adjacent* is the one left over (it's 'adjacent' to the angle, it touches it).

→ Sine, cosine and tangent

There are three commonly used terms in trigonometry – sine, cosine and tangent. I am not, in this book (remember that this topic is huge and could fill a book by itself), going to go into the full detail of everything they can do, but to give you an indication they can help you work out different lengths and angles in a triangle.

Take a triangle like above

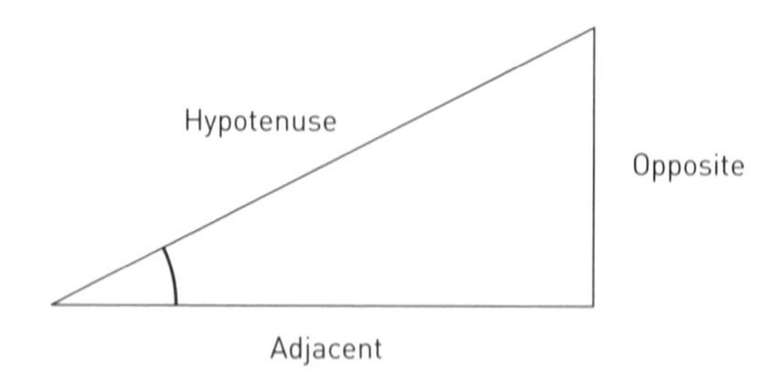

Then we define the following:

- **The *sine* of an angle is the length of the opposite side divided by the length of the hypotenuse.**
- **The *cosine* of an angle is the length of the adjacent side divided by the length of the hypotenuse.**
- **The *tangent* of an angle is the length of the opposite side divided by the length of the adjacent side.**

Here's an example:

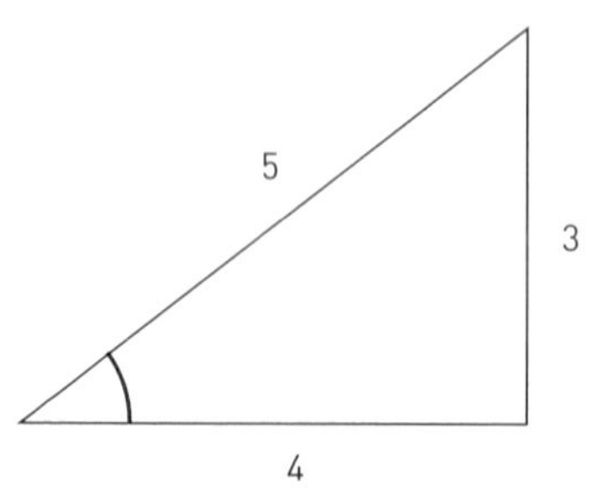

Then the sine of the angle is $\frac{3}{5}$, the cosine is $\frac{4}{5}$, and the tangent is $\frac{3}{4}$. Your calculator probably has buttons to do these – probably they are abbreviated as *sin*, *cos* and *tan*. Try them for yourself and work out the sine, cosine and tangent of an angle of your choice (say 20°). Note that you will almost always get complicated decimal answers

– can you find any angles where you actually get nice normal numbers?

> **smart tip**
>
> There are ways to learn things like this. For example, many people are taught the abbreviation SOHCAHTOA which sounds like a word – this is short for **S**ine (**O**pposite divided by **H**ypotenuse), **C**osine (**A**djacent divided by **H**ypotenuse) **T**angent (**O**pposite divided by **A**djacent). Use such tricks to help you learn things – for example you can learn the first few digits of π (3.1415) by learning a phrase where each word has the corresponding number of letters in it, such as 'hey, I want a pizza' where the words are of length 3, 1, 4, 1, 5 giving you a nice way to remember π this far. These sort of tricks don't suit everybody but they do some, and such techniques and little tricks can really help you memorise things that otherwise are hard to learn.

Summary

This is a very brief introduction to a topic – why so brief? This topic is so extensive that I couldn't fit everything you might need to know into a book, never mind a single chapter in such an introductory book. So I just want to introduce you to the idea.

Triangles are very important, for example in construction – you need to work out how long you need your bridges and beams to be. Sines, cosines and tangents can be really useful in helping work things out – just as an example, I will tell you that you can work out the area of a general triangle (when you just know a couple of sides and an angle) using sine.

This really is a topic where it is worth knowing the basics, but if you need more, then refer to a slightly more advanced textbook.

> **smart tip**
>
> This is an introductory textbook, and there are some topics that I can't cover in a lot of detail – this is deliberate. There is a wealth of further information out there – if you need to delve deep into a particular topic, then look at more advanced texts that focus on the particular thing you want to learn. It's unlikely that any single book will cover everything you need (if it does, it's probably going to be too big to even carry!) so focus your attention on further material that develops what *you* need to know in a particular area.

Exercises

1 What is the hypotenuse of the following triangles?

Example: Other side lengths 4 and 5

Solution: The hypotenuse is $\sqrt{4^2 + 5^2} = \sqrt{16 + 25} = \sqrt{41} = 6.403$ (3 decimal places).

(a) Sides 1 and 1 (b) Sides 3 and 7

(c) Sides 6 and 8 (d) Sides 5 and 12

2 If you know two angles in a triangle as given, what is the other angle?

Example: Two angles are 30° and 50°

Solution: Since the angles in a triangle make 180°, the other angle is 180° − 30° − 50° = 100°

(a) 10° and 30° (b) 100° and 70° (c) 89° and 89°

(Extra – can you draw a sketch of what these triangles look like?)

Why can't you have a triangle with two angles of 100° each?

3 Write down (as a fraction) the sine, cosine and tangent of the angle in the following triangle:

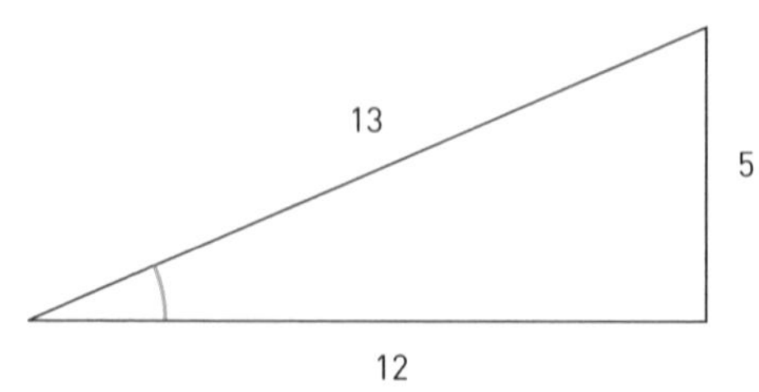

4 Use your calculator to obtain the sine, cosine and tangent of:

(a) 10° (b) 30° (c) 45° (d) 60°

(e) 90° (something will probably go wrong in this one – any idea why?)

Give your answers to 3 decimal places where appropriate.

5 *Extra – just for fun.*

As we mentioned in Chapter 5, π to more decimal places is 3.141592 65358979323846264338327950288419716939937510 ...

Can you come up with a phrase like I gave you before ('hey, I want a pizza' in the tip) to help you remember this to more decimal places?

21 Basic number theory

Some concepts in numbers that you might come across

In this very brief chapter I'll just run through a few terms and ideas that you might encounter in your studies, so you know what they mean. I'm not going to delve deeply into any of these – I just want to mention the ideas so that you can refer to this if you ever come across them!

Key topics

→ Classes of numbers
→ Prime numbers and factors
→ Factorials

Key terms
natural numbers integers real numbers prime factor decomposition coprime factorial

smart tip

It's worth seeing things, even if you don't use them often, so that they stay in your mind. You might never need to use these ideas, but if you do, then hopefully you can remember 'oh yes, that was in that book' and know where to go to read about it. Nobody expects you to learn and remember everything – but if you see something and remember having seen it, then you'll know where to look if you do need it – this is true for everything you study. There is no such thing as a completely useless piece of knowledge!

→ Types of numbers

The numbers that you know fall into a range of 'classes', which we'll describe below:

(a) The 'normal' numbers that you use all the time (0, 1, 2, 3, . . .) are called *whole numbers* or, more mathematically, referred to as *natural numbers.* (Note that there is sometimes an argument as to whether 0 is classed as a natural number or not – I say it is!)

These numbers are ones that are intuitively natural to us (hence the name). It might interest you to know that in Roman times (you may know Roman numerals?) there was no Roman numeral for 0, and negative numbers and fractions didn't come along until much later. The Romans didn't have the concept of zero and negatives, etc. – whilst this is natural to us now, it demonstrates the evolution of knowledge!

(b) If you include negative numbers as well, so ..., -3, -2, -1, 0, 1, 2, 3, ... this is referred to as *integers*.

(c) If you also include fractions, so things such as $\frac{1}{2}$ or $\frac{5}{7}$ then these are referred to as *rational numbers.*

(d) There are some numbers you can't actually write as fractions – examples include π and $\sqrt{2}$. These are called *irrational numbers*.

(e) All numbers that you can think of, including integers, fractions, irrational numbers, etc., all together are called *real numbers*.

→ Prime numbers and factors

A *factor* of a number is a number that divides exactly into it. For example, 3 is a factor of 12 since it divides exactly into 12.

You can list all the factors of 12 – they are 1, 2, 3, 4, 6 and 12 – all of these divide exactly into 12 with no remainder (note that we include 1 and 12).

A *prime number* is one that has no factors apart from 1 and itself. 13 is an example of a prime number – nothing divides into 13 apart from 1 and 13.

The first few prime numbers are 2, 3, 5, 7, 11, 13, 17, 19, 23, ... Maybe you can continue this yourself for a little while?

Note that we don't count 1 as a prime number. There are good reasons for this (I'll mention one reason shortly) but really just learn it as a rule.

Given any number, you can 'break it down' into prime numbers. For example, 30 can be written as $2 \times 3 \times 5$. It might surprise you to know that there is basically only one way to break a number down like this into prime numbers (OK, I could write $3 \times 5 \times 2$, but that's the same thing in a different order – there's no really different way). This is called the *prime factorisation* or *decomposition* of a number.

Note that while every number in a decomposition is prime, it's perfectly OK to have the same one more than once – for example $18 = 2 \times 3 \times 3$.

Something called the *Fundamental Theorem of Arithmetic* says that you can only break a number down like this (into prime numbers) in one way – note that if we allowed 1 to be a prime number I could break down 30 as $1 \times 2 \times 3 \times 5$, or $1 \times 1 \times 2 \times 3 \times 5$, etc. – so more than one way – a good reason not to count 1 as prime.

In a sense, prime numbers are the 'building blocks' of all numbers. Even so, we don't know much about them – we don't even know if there is a pattern to them (so a formula that would tell us straight away if a number is prime or not). If you could find such a formula, you could make yourself very rich!

Two numbers are said to be *coprime* if they have no factors in common apart from 1. So, for example, 9 and 11 are coprime as nothing divides into them both apart from 1, but 9 and 12 are not coprime as they both have 3 as a factor.

Such concepts (prime numbers, decompositions, etc) are used strongly in cryptography (keeping data secret on a computer) – every time you send an email or access a webpage, there is a lot of security going on, much of which relies on prime numbers – bear that in mind every time you use your computer!

→ Factorials

The *factorial* of a number appears quite commonly in maths. I'll give an example to define it: the factorial of 5, which is written as 5! (yes, really an exclamation mark) is defined as $5 \times 4 \times 3 \times 2 \times 1$, so all the positive integers up to 5 multiplied together, which you can work out to be 120.

Similarly $6! = 6 \times 5 \times 4 \times 3 \times 2 \times 1$ and so on.

Your calculator probably has a factorial button – can you find it?

Note that you can't really define a factorial for anything other than a positive integer – it doesn't make sense to talk about $(-3)!$ for example. Note, though, that we do define $0! = 1$ (the reasons are similar to why we define anything to the power 0 to be 1).

> **smart tip**
>
> I mentioned above that these concepts are used heavily in cryptography. That means that every time you send an email or load up a webpage, the security checking going on in the background is probably using prime numbers and maths. Do try to gain an awareness of how maths is all around you, and you'll get a much better appreciation as to why it is important!

Summary

This is just a very short look at a few topics in numbers that you might come across. Obviously in a book like this, I can't cover the whole of maths – but what I can do is demonstrate a small idea, and hopefully make you see how something relatively simple (such as a prime number) is actually really important. Maybe you will be inspired to find out something more and learn more about other topics not covered in this book – just always remember (for example, topics like this are used every time you turn on your computer) that maths is all around you!

Exercises

1 Write down all the factors of the following numbers. Are they prime or not?

Example: 12

Solution: Factors are anything that divides exactly into 12, so they are 1, 2, 3, 4, 6, 12. This is not prime as it has other factors apart from 1 and itself (12).

(a) 8	(b) 10	(c) 13	(d) 18
(e) 23	(f) 24	(g) 5	(h) 30

2 Write down the decomposition into primes of the following numbers.

Example: 42

Solution: $42 = 2 \times 3 \times 7$.

(a) 12 (b) 18 (c) 25 (d) 100

3 Calculate the following factorials.

Example: 5!

Solution: $5! = 5 \times 4 \times 3 \times 2 \times 1 = 120$.

(a) 3! (b) 4! (c) 6! (d) 10! (e) 0!

4 *Extra question – just for fun.*

Write down all the prime numbers up to 100. Then go onto the internet and find the largest prime number currently known to mankind – how many digits does it have?

22 Inequalities and basic logic

Using mathematical symbols to represent inequalities and basic logical concepts

A lot of maths is about 'equality' – one thing being equal to something else. There are other basic concepts in mathematics which are very important, which describe whether one thing is less than another, more than another, and so on.

Key topics

→ Inequalities
→ Logic
→ Implication

Key terms

inequality less than greater than equal to not equal to less than or equal to greater than or equal to logic 'or' 'and' 'not' implies

→ Inequalities

The intuitive idea of 'less than' and 'greater than' (or 'more than') is not too difficult to comprehend. Clearly 2 is 'less than' 3, and 20 is 'greater than' 10.

It is important to take care when dealing with these with negative numbers.

Remembering that the integers can be written as a 'number line'

$$\ldots \quad -6 \quad -5 \quad -4 \quad -3 \quad -2 \quad -1 \quad 0 \quad 1 \quad 2 \quad 3 \quad 4 \quad 5 \quad 6 \quad \ldots$$

you can see that -6 is smaller than -4 for example. Think of it in terms of temperatures – hotter is 'bigger' and colder is 'smaller'. -6 is colder than -4, so -6 is smaller than -4. Use this number line to help you with inequalities – smaller numbers are to the left of this number line.

You need to know the following symbols and what they mean – we will illustrate with several examples afterwards.

Symbol	*Meaning*
$<$	Less than
$\leqslant$	Less than **OR** equal to
$>$	Greater than
$\geq$	Greater than **OR** equal to
$=$	Equal to
$\neq$	**NOT** equal to

Examples

'Equal to' and 'not equal to' are fairly easy to understand. Two numbers are equal only when they are exactly the same. For example, $3 = 3$, $-4 = -4$, $0 = 0$, etc. On the other hand, almost every pair of numbers (in fact, every pair which aren't the same) are 'not equal'. $1 \neq 2$ for example, since 1 is not equal to 2. Similarly, $1 \neq 3$, $1 \neq 4$, $1 \neq 0$, $1 \neq -1$ and so on.

'Less than' means smaller than. So, for example, $2 < 3$ and $3 < 6$. Be careful with negative numbers. Remember the 'colder' analogy and look at the number line. You can see that, for example, $-6 < -4$, we have $-3 < -1$ and so on. Also $-2 < 0$, $-2 < 1$, etc. Just look at the number line.

For 'less than', we aren't allowed to have the numbers the same. It's not true to say '3 is less than 3', 3 isn't smaller than 3! With 'less than or equal to' we do allow this. So for example we can say $3 \leqslant 3$. 'Less than or equal to' means exactly that – it's either less than, or equal to. 3 might not be less than 3, but it is equal to it, so we are OK.

Similarly $2 \leqslant 3$, certainly they aren't equal but 2 is less than 3 so we're still OK.

Other examples of $\leqslant$ could be $0 \leqslant 4$, $-2 \leqslant 1$, $-6 \leqslant -3$, $-1 \leqslant -1$, $0 \leqslant 0$ and so on.

Remember:

$\leqslant$ means 'less than' OR 'equal to' – if either one is true, we're OK.

'Greater than' works the same way, just we talk about bigger rather than smaller. So $3 > 2$ because 3 is greater than 2. Similarly $5 > 1$, $1 > -2$, $-1 > -3$ and so on – again just look at the number line to see which one is bigger.

Similarly, 'greater than or equal to' means either greater than, or equal to, so now we can say things like $5 \geqslant 5$. Similarly, we have $5 \geqslant 3$, $2 \geqslant -1$, $1 \geqslant -4$, $-2 \geqslant -5$ and so on.

Note that the difference between 'greater than' and 'greater than or equal to' can be quite important. As a real-life example, you are eligible to vote in the UK if you are at least 18. That means your age is 'greater than or equal to' 18. If we used 'greater than', then someone of age 18 couldn't vote – their age is not greater than 18. So it does make an important difference.

smart tip

This sort of concept is entirely natural. When you watch the weather on TV, look to see which cities are warmer or colder than others (especially in winter months when some temperatures are negative). Also, think about the pass requirements on exams – say you have to get at least 40% to pass, that means your mark has to be greater than or equal to 40%. Whenever you watch an age-restricted film or purchase age-restricted products, an inequality is in place (your age is greater than or equal to 18, say). There are many examples in daily life of inequalities (cook for at least 40 minutes is another example) – think of as many as you can and realise how often you use this idea of inequality!

→ Basic logical concepts

You saw above the idea of 'greater than OR equal to' – this meant that the statement was true if the first number was either greater than the second, OR equal to it.

'Or' is an example of a commonly used word in English that can be applied to two statements – and it means either the first one is true, or the second one is true.

For example – suppose you are eligible for half-price travel if you are either under 16, or at least 65. This is an example of us using the word 'OR'. You could write this slightly more mathematically as

$$(\text{age} < 16) \text{ OR } (\text{age} \geq 65)$$

Note the use of 'less than' and 'greater than or equal to' in this example – again (as in the age for voting above) why is it so important to get this right?

Another commonly used word is 'AND'. For example, you might be eligible for a discount travel card if you are aged between 16 and 25. That means, your age is at least 16, and it's no more than 25. So you could again write this, slightly more mathematically, as

$$(\text{age} \geqslant 16) \text{ AND } (\text{age} \leqslant 25)$$

Note the important difference between OR and AND. With OR, only one of the two cases has to be true. With AND, they both have to be true.

For example, if you are told that you have to pass Test 1 **and** Test 2, then you know you need to pass them both. If you are told you have to pass Test 1 **or** Test 2 then you only need to pass one of them.

You use these words all the time in daily life, so don't be daunted; all we're doing is using them in exactly the same way here.

One other word you often use is 'NOT', which essentially means the opposite – the opposite of 'it is raining' is 'it is not raining'. For example, you might be allowed into a bar if you are at least 18. This is the same as saying 'you are not younger than 18' so the two statements

- $(\text{age} \geq 18)$
- NOT $(\text{age} < 18)$

are exactly the same thing. You've also seen the idea of 'NOT' used before with the symbol $\neq$ (short for 'not equal to') which is the 'opposite' of equals $=$.

Remember, these three common words OR, AND, NOT are being used exactly as in English.

In maths, we often use symbols to stand for things – you've seen this in algebra.

In this chapter we've already used symbols to stand for things we'd say in English as 'greater than', 'not equal to' and so on.

The words 'OR', 'AND' and 'NOT' also have symbols standing for them, which look like this:

Symbol	Meaning
$\wedge$	AND
$\vee$	OR
$\neg$	NOT

So, I could for example have written my examples above as

- $(\text{age} < 16) \vee (\text{age} \geq 65)$ (age is less than 16, *or* greater than or equal to 65).
- $(\text{age} \geq 16) \wedge (\text{age} \leqslant 25)$ (age is greater than or equal to 16 *and* less than or equal to 25)
- $\neg\,(\text{age} < 18)$ (age is *not* less than 18).

Let's have a look at a few examples purely using numbers.

Example

Are the following statements true or false?

(i) $(3 < 3) \wedge (2 < 4)$

(ii) $(3 < 3) \vee (2 < 4)$

(iii) $\neg\,(2 < 3)$

For (i), the symbol $\wedge$ means 'AND'. So, the statement reads as 'is it true that both $3 < 3$, and $2 < 4$'? Well, $2 < 4$ so that's fine, but it's not true that $3 < 3$, so this statement is false, the two parts aren't BOTH correct.

For $\wedge$ 'AND', both things have to be true.

For (ii), the symbol $\vee$ means 'OR'. So the statement reads as 'is it true that either
$3 < 3$, or $2 < 4$'? Well, $3 < 3$ is not true, but the other one $(2 < 4)$ is fine, and so this is indeed true – either one or the other is true, and so we're fine.

For $\vee$ 'OR', only one of the things has to be true.

(It's perfectly fine if both are true).

For (iii), the symbol $\neg$ means 'NOT', so take the opposite answer.
$2 < 3$ is a true statement, and so $\neg\,(2 < 3)$ is the opposite, i.e false.

$\neg$ 'NOT' is the opposite answer.

Hopefully you're seeing that the use of these symbols is just to make life a bit quicker. Instead of writing down words we're writing down a 'shorthand symbol'. It is much quicker to write $(3 < 3) \wedge (2 < 4)$ rather than *'is it true that three is both less than three, and also that two is less than four?'* You've already seen a lot more about the use of symbols to make things shorter when we did algebra previously.

→ Implication

As a final note on 'logic', we'll briefly consider the idea of implication.

The word 'implication' means 'following from'. For example, suppose it is raining and you look outside onto your garden. From the fact that it's raining, you can deduce that your grass is going to be wet. Hence 'it is raining' implies 'the grass is wet'.

Does it follow the other way round? If you look outside and the grass is wet, does that mean it's raining? Not necessarily, somebody might have been watering it with a hose. Hence 'the grass is wet' does NOT imply 'it is raining'.

Similarly, you know that all coal is black. So 'something being coal' implies 'that thing is black'. But the other way round isn't true, 'something is black' doesn't necessarily mean it's coal – it could be lots of things (a car, a vinyl record, a dog ...).

The mathematical symbol for 'implies' is $\Rightarrow$. You don't need to know anything about this for now, but it's worth thinking about how what we are doing is just using short-hand symbols to symbolise normal English sentences.

Also, think about how we are deducing things from something else, but we have to be careful not to jump to conclusions going the other way round. You can imagine how important it is to get this right, and deduce the right conclusion, in almost any academic subject (medicine, policing, law, etc.).

Think about how many times in your day you use a word like 'or', 'and', 'not'. On each occasion, what would happen if you got them wrong? For example, would you be happy in a cafe if you ordered a meal and a drink and they used 'or' and just brought you one of the two things? Or if you told someone 'not to do something' and they ignored the 'not' and did it? Of course not. Just always remember that we are using these maths symbols exactly as in daily life – realising how much you are using these ideas makes you realise that maths plays a much bigger part of your life than you ever thought!

Summary

This topic was deliberately chosen to conclude this book, as it demonstrates clearly the fact that mathematics is not meant to be something from another planet, but something genuinely useful that is relevant to our world and our daily life. Inequalities occur everywhere (you must drive at a speed less than or equal to 30mph, say) and the maths is just there to help write it down – surely you agree that it is easier to write $3 \leq 4$ rather than to write out 'the number three is less than or equal to the number 4'?

Maths is meant to help – hopefully with a topic like this, you can appreciate that it is real life and really relevant.

Exercises

1 For each of the following, state whether they are true or false.

Example: $2 \geq 5$

Solution: 2 is neither greater than 5, or equal to it, so this statement is false.

(a) $3 < 4$ (b) $2 \leqslant 9$ (c) $5 < 3$ (d) $3 \leqslant -2$

(e) $-2 \leqslant -3$ (f) $-7 < -2$ (g) $4 < 4$ (h) $3 > 9$

(i) $3 \geqslant -2$ (j) $2 \geqslant -9$ (k) $2 \geqslant 2$ (l) $-2 > 2$

(m) $0 > 0$ (n) $3 = 4$ (o) $2 \neq -2$ (p) $4 \neq 4$

2 For each of the following, state whether they are true or false:

Example: $(3 < 2) \vee (4 \geqslant 1)$

Solution: The first of these $(3 < 2)$ is false, but the second one is true $(4 \geqslant 1)$. As the symbol in the middle is an OR $\vee$, then because at least one of the two is true, then the overall statement is true.

(a) $(4 > 3) \wedge (2 < 3)$

(b) $(3 > 2) \wedge (3 < 3)$

(c) $(4 \geqslant 3) \wedge (3 < 3)$

(d) $(4 > 3) \vee (3 < 2)$

(e) $(2 > 3) \vee (3 < 3)$

(f) $(4 \geqslant 3) \vee (3 \geqslant 3)$

(g) $(-4 > -3) \wedge (-2 < -3)$

(h) $(-2 > -3) \wedge (-4 < -2)$

(i) $(-2 \geqslant -4) \wedge (3 \geq 3)$

(j) $(-4 > -3) \vee (-2 < -3)$

(k) $(-2 > -3) \vee (-4 < -2)$

(l) $(-2 \geqslant -4) \vee (3 \geqslant 3)$

(m) $\neg(2 < 3)$

(n) $\neg(2 \geqslant 2)$

(o) $\neg(4 \neq 4)$

3 Given the following pairs of statements, which way round does the implication go (which statement follows from which)? Why does it not follow the other way round?

(a) I can't see the sun.
It is very cloudy today.

(b) The man was murdered.
The man is dead.

(c) The patient needs some medicine.
The patient has a very high temperature.

(d) The player is unfit for this match and can't play.
The player won't score any goals in this match.

Summary

The main hope from writing this book is that it has given you an appreciation of the role mathematics has to play in your daily life and studies, and has given you some confidence to be able to say that you actually 'can do maths'.

Always try to keep in your mind the fact that you use maths all the time in your life without thinking about it. Every time you go shopping, plan a journey, read statistics in a newspaper or analyse a problem, essentially you are using mathematics.

Mathematics really is part of your life. If nothing else, I hope that this book has opened your eyes to that fact and given you some confidence.

Whatever course you are doing and whatever path you follow in life, you will be using mathematics even if you didn't really realise it. Maybe sometimes when you are doing something you will recognise the maths that you are actually using – hopefully this book will inspire that.

It only remains for me to wish you good luck in your studies and your future!

Further work on our website

Please don't forget that there is much more available for you on the website for this book, which you find by linking through from the Smarter Student website at

http://www.smartstudyskills.com

On the site you can find full worked solutions to all the questions, together with further questions (some related to particular fields of study) for you to practise with, further links, contact details, and plenty more, including some fun challenges – so do visit us there.

Glossary

The following is a list of the most important terms used in the book.

Adjacent side the shorter of the two sides forming a given angle – so the third side in a triangle after the hypotenuse and opposite side.

Algebra the branch of mathematics involving the use of symbols to stand for numbers.

Area a measure of what is contained inside a two-dimensional shape.

Arithmetic the study of numbers.

Axis either of the horizontal or vertical lines used to define a graph.

Bar chart a type of graph where values are represented as bars.

Base the 'main number' in an expression like 2^8, which has base 2. Also used in logarithms, for example $\log_2(8)$ has base 2.

BODMAS the rule that says you should do brackets first in any calculation, then divisions and multiplications, then finally additions and subtraction.

Circumference the distance all around a circle.

Coefficient the 'main number' in an expression written in scientific notation, for example the coefficient in 3×10^8 is 3. Also used for the number in front of a symbol in algebra, for example the coefficient of $5x$ is 5.

Collecting like terms putting together terms in an algebraic expression that have the same symbols in them, for example $4x + 7x = 11x$.

Commutative the property that it doesn't matter what order you do something – for example addition is commutative since, for example $3 + 4 = 4 + 3$.

Complement the opposite of something, often used in probability, for example the complement of choosing an odd number is choosing an even number.

Cone a three-dimensional shape with a circular base tapering to a point (like a traffic cone).

Coprime having no factor (apart from 1) in common.

Correlation a relationship between two things.

Cosine for an angle, the adjacent side divided by the hypotenuse.

Cube a cuboid where all the sides are of equal length. Also used to refer to a power of 3.

Cuboid a three-dimensional shape where all the sides are rectangles (like a brick).

Cylinder a three-dimensional shape consisting of a circular tube (like a tin).

Decimal places the number of digits after a decimal point.

Decomposition breaking a number down into a multiplication of prime numbers such as $30 = 2 \times 3 \times 5$.

Denominator the bottom part of a fraction.

Depth the measure of how 'deep' a three-dimensional shape is.

Diameter the distance across a circle (going through the middle) – equal to twice the radius.

Disjoint any two events that have nothing in common with each other.

Dispersion a measure of how far a set of values are spread away from their 'average'.

Distributive – the property that $x(y + z) = xy + xz$.

e a special number in mathematics, roughly equal to 2.718281828...

Evaluation the process of putting particular values into an algebraic expression.

Even a number that is a multiple of 2, such as 2, 4, 6, 8 etc.

Event a specific possible outcome, such as 'heads' when you toss a coin.

Expanding the brackets removing the brackets from an expression, such as writing $xy + xz$ rather than $x(y + z)$.

Exponent the 'power of 10' in an expression written in scientific notation, for example the exponent in 3×10^8 is 8. Also used in general for any power.

Factor a number that divides exactly into another, for example 6 is a factor of 18.

Factorial the product of all positive integers up to a given number, written using the symbol ! (exclamation mark), for example $5! = 5 \times 4 \times 3 \times 2 \times 1 = 120$.

Factorisation the reverse of expanding brackets, such as writing $x(y + z)$ rather than $xy + xz$ or writing $(x + 2)(x + 3)$ rather than $x^2 + 5x + 6$.

Fundamental theorem of arithmetic the fact that any number can only be broken down into a product of primes in only one basic way.

Googol the number 10^{100} (1 followed by 100 zeros).

Gradient a measure of the slope of a line, often abbreviated as m in calculations.

Graph a pictorial representation of data.

Height the distance 'up' a shape.

Hypotenuse the longest side in a right-angled triangle.

Improper fraction another name for a top-heavy fraction.

Index the number of times a number is multiplied together in an expression such as 2^8, which has index 8.

Indices the plural of index.

Inequality any mathematical expression involving things that are not equal (including 'less than', 'greater than', and so on).

Integer any number ...., −3, −2, −1, 0, 1, 2, 3, ...

Intercept the point at which a straight line cuts the y-axis, often referred to as c in calculations.

Law a defined mathematical formula that is always true.

Line graph a connection of points on a graph, obtained by drawing a series of lines between them.

Linear equation any equation with only single powers of x (or other symbols).

Logarithm the question of asking what power a number has to put to, to get another number, for example $\log_2(8) = 3$ since $2^3 = 8$.

Lowest common multiple the smallest whole number that two numbers both divide into, for example the lowest common multiple of 12 and 8 is 24.

Lowest terms writing a fraction so that it has been cancelled down as much as possible.

Mean the average of a set of values, calculated as the sum of the values divided by the number of values.

Median the 'middle' of a set of values, calculated by listing them in order and choosing the middle one (or the mean of the middle two if there is no middle element).

Mixed fraction a fraction such as $2\frac{2}{3}$ which contains a whole number part and a fractional part.

Mode the most common value in a list of values.

Natural logarithm a logarithm to base e.

Natural number any number 0, 1, 2, 3, 4, 5 ...

Negative number a number less than 0.

Numerator the top part of a fraction.

Odd a number that is not a multiple of 2, such as 1, 3, 5, 7, etc.

Opposite side the side opposite an angle.

Parallelogram a four-sided shape where opposite sides are parallel (go in the same direction).

Percentage a fraction with denominator 100, often abbreviated to % (so $\frac{37}{100}$ = 37% for example).

Pi a special number in mathematics, denoted π, roughly equal to 3.14159265..., often taken to be 3.142 for calculations.

Pie chart a type of graph where values are represented as segments of a circle.

Polynomial an expression involving powers like x^2, x^3, etc.

Positive number a number greater than 0.

Power another word for index.

Prime number a number that is only divisible by 1 and itself.

Probability the study of chances and the likelihood of something happening.

Product another word for multiplication.

Proportion the percentage of one thing inside a whole set – for example the proportion of boys in a class of 8 boys and 12 girls is 40% (8 out of 20).

Proportionate two things that have exactly the same proportion.

Protractor a mathematical tool to help you draw angles accurately.

Pythagoras' theorem the theorem that states that the square of the hypotenuse is equal to the sum of the squares of the other two sides in a right-angled triangle.

Quadratic equation an equation involving squared powers like x^2.

Quadratic formula the formula $\frac{-b \pm \sqrt{b^2 - 4ac}}{2a}$ used to solve quadratic equations.

Quotient how many times one number divides into another, for example the quotient when you divide 6 into 20 is 3 (since three 6s go into 20 (with remainder 2)).

Radius the distance from the centre of a circle to the edge.

Range the difference between the largest and smallest values in a set of data.

Ratio a relationship between two quantities detailing 'how many' of the second quantity compared with the first quantity – for example if there are 6 boys and 10 girls the ratio is 3:5 (three boys for every five girls).

Real number any number that you naturally know, including square roots, fractions, π, and so on.

Rectangle a four-sided shape with four right angles.

Remainder what is left over when you divide one number into another – for example the remainder when you divide 20 by 6 is 2, since 6 divides into 18 and there is 2 left over.

Right angle an angle that is 90°, or alternatively created by two perpendicular lines.

Right-angled triangle a triangle where one of the angles is a right angle.

Scatter graph a graph created by plotting a set of points, with no lines between them.

Scientific notation writing a number as a number multiplied by a power of 10, for example 3×10^8.

Segment a part of a circle.

Semicircle half of a circle.

Significant figures the number of relevant non-zero digits in an approximation (for example 12000 has two significant figures) – don't confuse with decimal places.

Simplifying making an algebraic expression as simple as possible.

Simultaneous equations two (or more) linear equations that need to be solved at the same time.

Sine for an angle, the opposite side divided by the hypotenuse.

SOHCAHTOA an acronym to help remember the sine, cosine and tangent definitions (sine = opposite/hypotenuse, cosine = adjacent/hypotenuse, tangent = opposite/adjacent).

Sphere a three-dimensional circle (like a football).

Square (meaning 1) to multiply a number by itself, or to put it the power 2, such as 6^2 which is 36.

Square (meaning 2) a rectangle where all four sides are of equal length.

Square root a number that when you multiply it by itself gives a given number – for example the square root of 36 is 6 since $6 \times 6 = 36$.

Standard deviation a measure of dispersion equal to the square root of the variance.

Statistics the study of data and its analysis.

Subject a single symbol on the left-hand side of an equation, for example in the equation $x = y + z$, the subject is x.

Tangent for an angle, the opposite side divided by the adjacent side.

Term any of the things added together in an algebraic expression.

Times tables a list of multiplications of numbers, such as a list of all multiplications of numbers up to 12.

Top-heavy fraction a fraction where the numerator (top) is greater than the denominator (bottom).

Transposition the act of moving things around in algebraic expressions.

Triangle a shape made up of three straight lines.

Trigonometry the study of triangles.

Variance a measure of dispersion, obtained by adding together the squares of all the differences between the values and the mean, and then dividing by the number of values.

Volume a measure of what is contained inside a three-dimensional shape.

Width the distance 'across' a shape.

Some useful things

The following is a collection of the most useful things in the book (formulae and so on) together with a few bits of extra information.

Names of numbers

Thousand = 10^3 = 1000
Million = 10^6 = 1000000
Billion = 10^9 = 1000000000
Trillion = 10^{12} = 1000000000000

Googol = 10^{100}

Tenth = 10^{-1} = 0.1
Hundredth = 10^{-2} = 0.01
Thousandth = 10^{-3} = 0.001

Units of measurement

The standard unit of length is the metre – abbreviated to m. The following are also common:

kilometre (km) = 1000m
centimetre (cm) = 0.01m (100 centimetres in a metre)
millimetre (mm) = 0.001m (1000 millimetres in a metre, also 10 millimetres in a centimetre)

Old-fashioned measurements of length include:

inch = approximately 2.54cm
foot = 12 inches
yard = 3 feet (a bit less than a metre)
mile = 1760 yards, which is around 1609m

Temperature is measured in *Celsius* where 0°C is where water turns to ice, and 100°C is where water boils. Typical temperatures in the UK range from around −10°C in the deepest winter to 30°C in a hot summer, although it does occasionally get colder or hotter than this.

An old-fashioned way to measure temperature is *Fahrenheit*. You can convert from Celsius to Fahrenheit (f) by using the formula $f = \frac{9c}{5} + 32$ where c is the temperature in Celsius.

Note that these older units are rapidly becoming replaced by the metric units (metres, Celsius, and units of weight like kilograms) and you should be completely familiar with them – although some older measurements like pints do still commonly exist.

The BODMAS rule

In calculations you always do **B**rackets first, followed by **M**ultiplications and **D**ivisions, and finally **A**dditions and **S**ubtractions.

Multiplying positive and negative numbers

- A positive number multiplied by a positive number gives a positive answer
- A positive number multiplied by a negative number gives a negative answer
- A negative number multiplied by a positive number gives a negative answer
- A negative number multiplied by a negative number gives a positive answer

The laws of indices

For any numbers x, m, n:

- $x^m x^n = x^{m+n}$
- $\frac{x^m}{x^n} = x^{m-n}$
- $(x^m)^n = x^{mn}$
- $x^0 = 1$ (apart from 0^0 which is undefined)
- $x^{-m} = \frac{1}{x^m}$
- $x^{1/n} = \sqrt[n]{x}$ (the nth root of x, so $x^{1/2} = \sqrt{x}$)

The laws of logarithms

- $\log_a(xy) = \log_a(x) + \log_a(y)$
- $\log_a(\frac{x}{y}) = \log_a(x) - \log_a(y)$
- $\log_a(x^y) = y \times \log_a(x)$

Measures of location

The quick way to define them:

Mean: add up the values, and divide by the number of values
Median: the value in the middle
Mode: the most common value

Areas and volumes:

- Area of a rectangle $= wh$ where w is the width and h is the height
- Area of a square $= w^2$ where w is the width
- Area of a paralellogram $= wh$ where w is the width and h is the vertical height

- Area of a triangle $= \frac{1}{2}wh$ where w is the width and h is the vertical height
- Area of a circle $= \pi r^2$ where r is the radius
- Area of a semicircle $= \frac{1}{2}\pi r^2$ where r is the radius

- Volume of a cuboid $= whd$ where w is the width, h is the height and d is the depth
- Volume of a cube $= w^3$ where w is the width
- Volume of a cylinder is $\pi r^2 h$ where r is the radius and h is the height
- Volume of a sphere is $\frac{4}{3}\pi r^3$
- Volume of a cone is $\frac{1}{3}\pi r^2 h$

The quadratic formula

The two solutions of the equation $ax^2 + bx + c = 0$ are given by

$$x = \frac{-b \pm \sqrt{b^2 - 4ac}}{2a}$$

Trigonometry

Pythagoras' theorem: in a right-angled triangle with hypotenuse h and other sides a and b, then you have $h^2 = a^2 + b^2$.

sine = opposite/hypotenuse
cosine = adjacent/hypotenuse
tangent = opposite/adjacent

Logical symbols

$=$	equal to
$\neq$	not equal to
$<$	less than
$\leqslant$	less than or equal to
$>$	greater than
$\geqslant$	greater than or equal to
$\wedge$	and
$\vee$	or
$\neg$	not

Equation of a straight line

$y = mx + c$ where m is the gradient and c is the intercept

Roman numerals

I = 1   V = 5   X = 10   L = 50   C = 100   D = 500   M = 1000

π to 1000 decimal places (OK, this one isn't so useful)

3.1415926535897932384626433832795028841971693993751058209749445923
0781640628620899862803482534211706798214808651328230664709384460

9550582231725359408128481117450284102701938521105559644622948954930381964428810975665933446128475648233786783165271201909145648566923460348610454326648213393607260249141273724587006606315588174881520920962829254091715364367892590360011330530548820466521384146951941511609433057270365759591953092186117381932611793105118548074462379962749567351885752724891227938183011949129833673362440656643086021394946395224737190702179860943702770539217176293176752384674818467669405132000568127145263560827785771342757789609173637178721468440901224953430146549585371050792279689258923542019956112129021960864034418159813629774771309960518707211349999998372978049951059731732816096318595024459455346908302642522308253344685035261931188171010003137838752886587533208381420617177669147303598253490428755468731159562863882353787593751957781857780532171226806613001927876611195909216420198

Solutions to exercises

Note: These are only the final answers. For fully worked solutions, plus additional questions, please see the website.

Chapter 1 Basic arithmetic and the BODMAS rule

1 (a) Both are 10 (b) Both are 30 (c) Both are 6 (d) Both are 60

2 (a) Quotient = 3, remainder = 4
(b) Quotient = 4, remainder = 2
(c) Quotient = 4, remainder = 5
(d) Quotient = 4, remainder = 0

3 (a) 11 (b) 15 (c) 5 (d) 22 (e) 28
(f) 9 (g) 6 (h) 7 (i) 11

4 (a) 42 (b) 8 (c) 49 (d) 28 (e) 31 (f) 39

5 (a) £31 (b) £3 (c) 15p
(d) BuyNow is cheaper (cost is £18 compared with MegaDeals' £20)

Chapter 2 Negative numbers

1 (a) -2 (b) 3 (c) 2 (d) -4
(e) -1 (f) -8 (g) -10 (h) -2
(i) 0 (j) 0 (k) -3 (l) -5

2 (a) -8 (b) -24 (c) 16 (d) -27 (e) 100
(f) -6 (g) -3 (h) 2 (i) 1

3 (a) -9 (b) -1 (c) -17 (d) 12 (e) 6 (f) -5

4 (a) (i) 10°C (ii) -10°C (iii) After 4 hours
(b) 6 hours

Chapter 3 Fractions

1 (a) Numerator is 3, denominator is 7
(b) Numerator is 8, denominator is 9
(c) Numerator is 11, denominator is 3

2 (a) $\frac{1}{2}$ (b) $\frac{1}{4}$ (c) $\frac{2}{3}$ (d) $\frac{4}{9}$ (e) $\frac{2}{25}$ (f) $\frac{7}{15}$

3 (a) $\frac{3}{8}$ (b) $\frac{2}{5}$ (c) $\frac{5}{6}$ (d) $\frac{1}{6}$ (e) 1 (f) $\frac{49}{12}$

4 (a) $\frac{2}{3}$ (b) $\frac{8}{9}$ (c) $\frac{3}{5}$ (d) $\frac{9}{8}$ (e) $\frac{1}{6}$ (f) $\frac{20}{27}$

5 (a) $\frac{7}{12}$ (b) $\frac{23}{21}$ (c) $\frac{17}{9}$ (d) $\frac{23}{20}$ (e) $\frac{5}{2}$ (f) 4

6 (a) $\frac{1}{12}$ (b) $\frac{2}{21}$ (c) $\frac{1}{2}$ (d) $\frac{41}{77}$ (e) $\frac{17}{15}$ (f) $\frac{11}{12}$

7 (a) There is exactly ¼ litre left so you can fill one more $\frac{1}{4}$ litre glass.
(b) $\frac{1}{3}$ GB

Chapter 4 Percentages, ratios and proportions

1 (a) 70% (b) 60% (c) 86% (d) 28%

2 (a) 60% (b) 55% (c) 70% (d) 75%

3 (a) $\frac{3}{5}$ (b) $\frac{1}{20}$ (c) $\frac{24}{25}$ (d) $\frac{33}{100}$

4 (a) 30 (b) 16 (c) 5 (d) 45

5 (a) 2:1 (b) 3:4

6 (a) 9 black and 3 red (b) 10 pennies and 15 pounds

7 (a) The first class (52%) has a higher proportion than the second class (50%)
(b) Both forums have the same proportion (70%)

Chapter 5 Decimals, decimal places and significant figures

1 (a) 0.7 (b) 0.6 (c) 0.25 (d) 0.06
(e) 0.022 (f) 5.5 (g) 4.6 (h) 4.25

2 (a) $\frac{3}{10}$ (b) $\frac{4}{5}$ (c) $\frac{1}{4}$ (d) $\frac{12}{25}$
(e) $2\frac{3}{10}$ or $\frac{23}{10}$ (f) $4\frac{1}{2}$ or $\frac{9}{2}$ (g) $\frac{33}{100}$ (h) $\frac{1}{100}$

3 (a) 5.386 (b) 4.626 (c) 0.079 (d) 0.197
(e) 2.893 (f) 3.080 (g) 2.290 (h) 0.800

4 (a) 23100 (b) 1290 (c) 220000 (d) 400000
(e) 1.25 (f) 12.9 (g) 0.380 (h) 0.000139

Chapter 6 Scientific notation

1 (a) 2×10^4 (b) 2.34×10^7 (c) 5.6×10^6 (d) 1.3×10^{11}
(e) 1.53×10^2 (f) 1×10^7 (g) 1.932486×10^3 (h) 7×10^0

2 (a) 3.6×10^{-3} (b) 7.9×10^{-5} (c) 9.99×10^{-6} (d) 1.78×10^{-3}
(e) 7.8×10^{-10} (f) 1×10^{-8} (g) 1×10^{-1} (h) 1.34×10^{-1}

3 (a) 8×10^{10} (b) 4×10^{18} (c) 6×10^4
(d) 8×10^{-6} (e) 3.5×10^{16} (f) 2.4×10^{-8}

4 (a) 2×10^8 (b) 4.5×10^5 (c) 3×10^{12}
(d) 1.25×10^{-11} (e) 2.5×10^6 (f) 1×10^1

5 (a) 80 seconds (b) 1.3×10^6 seconds

Chapter 7 Indices

1 (a) 8 (b) 100 (c) 49 (d) 27
(e) 64 (f) 1 (g) 1 (h) $\frac{1}{4}$
(i) $\frac{1}{81}$ (j) 2 (k) 8 (l) 2

2 (a) 2^9 (b) 8^{22} (c) 2^5 (d) 4^{-10}
(e) 5^6 (f) 4^5 (g) 6^{-8} (h) 2^5
(i) 2^{12} (j) 3^{-6} (k) 4^4 (l) $3^0 = 1$

3 No – answers will always be positive (there is a branch of mathematics called *complex numbers* which does allow negative answers, but that's far beyond what you need to know for now!).

Chapter 8 An introduction to algebra

Please note that if you have the same answer but in a different order (e.g. $9y + 5x$ rather than $5x + 9y$) then that is absolutely fine and correct.

1 (a) 6 (b) 2 (c) 8 (d) 2
(e) -2 (f) 16 (g) 12 (h) 1

2 (a) 22 (b) -2 (c) 18 (d) 4
(e) 6 (f) -23 (g) 5
(h) The denominator is 0, and you can't divide by 0.

3 (a) $13x$ (b) $-2x$ (c) x
(d) $5x + 9y$ (e) $3x + 7y$ (f) $-x - y$
(g) $-6x$ (h) $\frac{3x}{10}$ (i) $\frac{13x + 4}{4}$ (or $\frac{13x}{4} + 1$)
(j) $8xy - 2y + 2$ (k) $2xy$ (l) $2 + x + xy$

4 (a) (i) $4c + 7d$ where c is the number of CDs and d is the number of DVDs.
(ii) I spend £34 and my friend spends £36, so he spends more
(b) (i) $10 + 5p$ where p is the number of packages
(ii) £15 (iii) £40 (iv) £510

Chapter 9 Brackets in algebra

Note that if you have written the symbols in a term in a different order (e.g. yx rather than xy), then that is absolutely fine.

1 (a) Both 18 (b) Both 4 (c) Both 6 (d) Both -2 (e) Both -2 (f) Both 0

2 (a) $xy + xz$ (b) $ab + ac$ (c) $rs - rt$

(d) $2x + 2y$ (e) $4x - 4y$ (f) $4x + 6y$
(g) $2xy + 2xz$ (h) $3x^2 - 3xy$ (i) $x^2y + xy^2$
(j) $3x + xy$ (k) $zx - 2z$ (l) $2xy - 2yz$

3 (a) $5x + 5y$ (b) $5x$
(c) $-x + 5y$ (d) $4x - 3y$
(e) $x^2 + 3xy + y^2$ (f) $x^2 + y^2$

4 (a) $tr + ts + ur + us$ (b) $xz - xw + yz - yw$ (c) $2xz + 3xw + 4yz + 6yw$
(d) $2x^2 + 5xy + 2y^2$ (e) $2x^2 - 3xy - 2y^2$ (f) $4x^2 - z^2$

5 (a) $2(x + y)$ (b) $x(3y + 2z)$ (c) $3x(z + 2y)$
(d) $xy(z + w)$ (e) $5y\,(2x + 1)$ (f) $xy(x + y)$

Chapter 10 Solving linear equations

1 (a) 4 (b) -3 (c) 7 (d) 2
(e) 6 (f) $\frac{3}{7}$ (g) 12 (h) $\frac{1}{2}$

2 (a) 4 (b) 2 (c) 2
(d) -1 (e) 2 (f) $\frac{1}{2}$
(g) 5 (h) -1 (i) 0

3 (a) 4 (b) 6 (c) 6 (d) 4
(e) 5 (f) $-\frac{1}{2}$ (g) 0

Chapter 11 Transposition and algebraic fractions

If you have answers in a different order, e.g. $b + a$ rather than $a + b$, that is absolutely fine.

1 (a) $s = r + 2t$, $s = 10$ (b) $s = r - t$, $s = -2$
(c) $s = \frac{t - r}{2}$, $s = 1$ (d) $s = 2t - r$, $s = 6$
(e) $s = \frac{r - 3}{t}$, $s = -\frac{1}{4}$ (f) $s = \frac{rt + 1}{t}$, $s = \frac{9}{4}$

2 (a) $x = z + y$ (b) $x = z - w - y$ (c) $x = z - y$
(d) $x = y - z$ (e) $x = \frac{z - y}{2}$ (f) $x = \frac{3z - y}{2}$
(g) $x = \frac{z}{y}$ (h) $x = \frac{z + 1}{y}$ (i) $x = yz$
(j) $x = \frac{y}{z}$ (k) $x = y(w + z)$ (l) $x = \frac{z}{2(4w - 3y)}$

3) (a) $\frac{5x}{7}$ (b) $\frac{7x}{12}$ (c) $\frac{4x}{11}$
(d) $\frac{x}{4}$ (e) $\frac{y + x}{xy}$ (f) $\frac{4y + 3x}{xy}$
(g) $\frac{xz}{yw}$ (h) $\frac{x}{z}$ (i) 4

Chapter 12 Simultaneous equations

1 (a) $x = 4, y = 2$ (b) $x = 3, y = 2$ (c) $x = 3, y = 1$
(d) $x = 3, y = -4$ (e) $x = -1, y = -1$ (f) $x = 1, y = 1$

2 (a) posters £3 and CDs £5
(b) pears are 65p and apples 52p
(c) 44 and 38

Chapter 13 Presentation of data

1 (a)

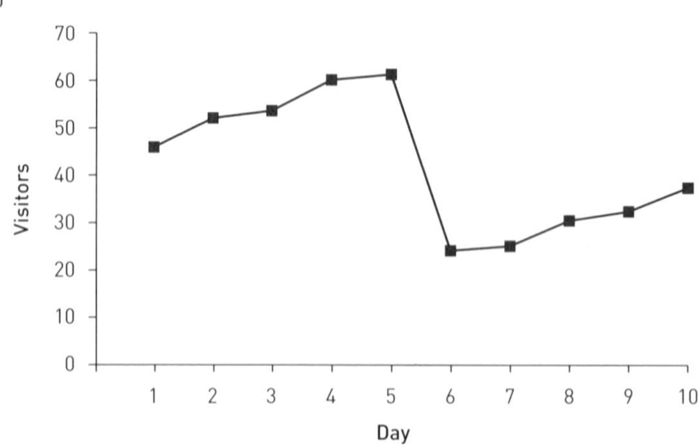

(b)

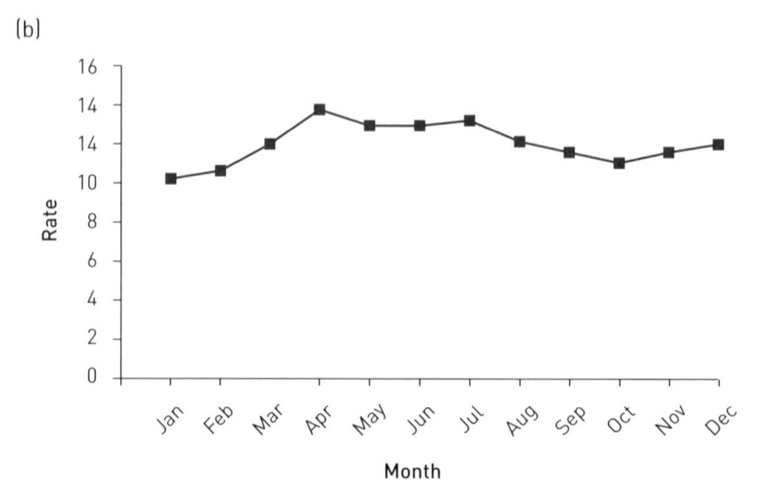

2 (a)

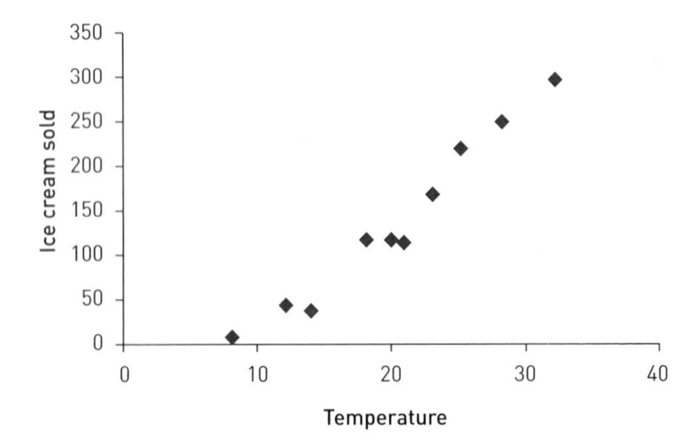

(b)

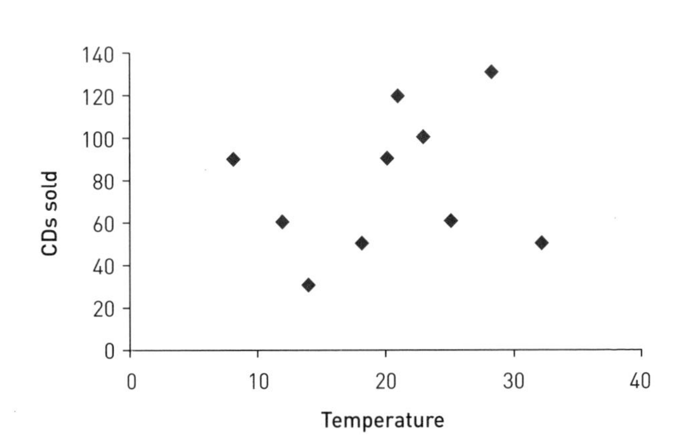

There appears to be a correlation with ice-creams but not with CDs.

3 (a)

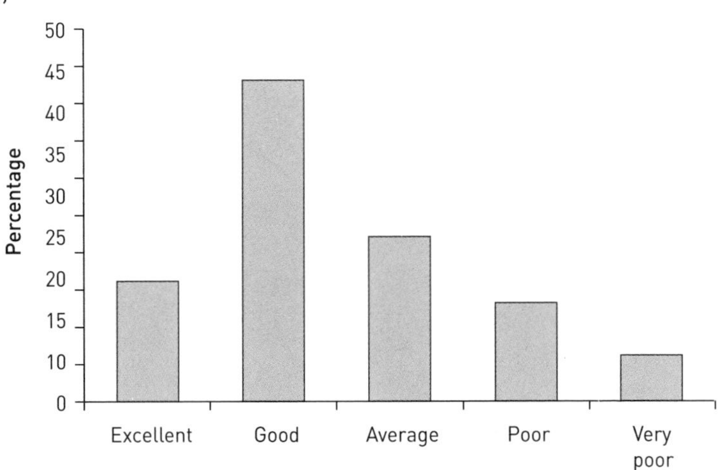

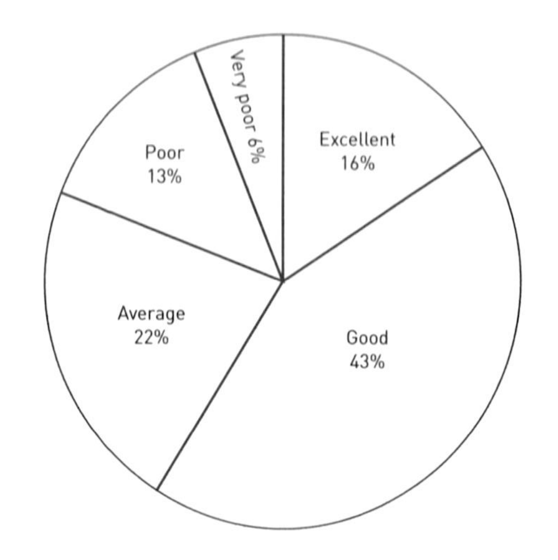
Very poor 6%
Excellent
16%
Poor
13%
Average
22%
Good
43%

(b)

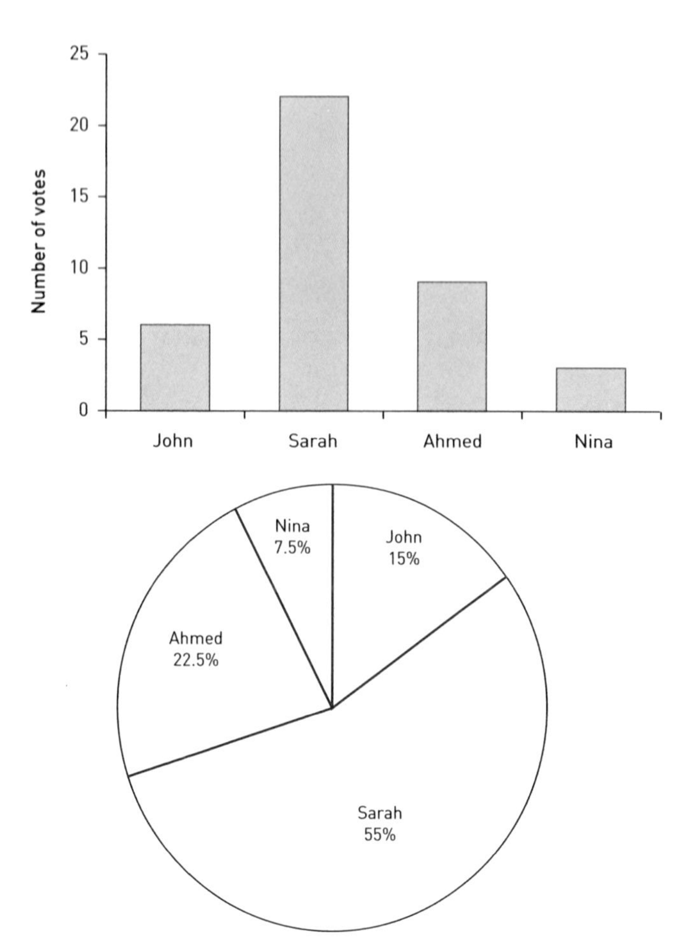
Number of votes
25
20
15
10
5
0
John
Sarah
Ahmed
Nina
Nina
7.5%
John
15%
Ahmed
22.5%
Sarah
55%

Chapter 14 Measures of location and dispersion

1 (a) 7	(b) 7	(c) 6	(d) 7	(e) 4.57	(f) 2.14
2 (a) 31	(b) 22	(c) 17	(d) 72	(e) 416.4	(f) 20.41
3 (a) 3	(b) 2.5	(c) 2 and 3	(d) 8	(e) 6	(f) 2.45

Chapter 15 Straight lines

As long as your graph is accurate, that is fine – you don't have to have exactly the same scale as given here.

1 (a)

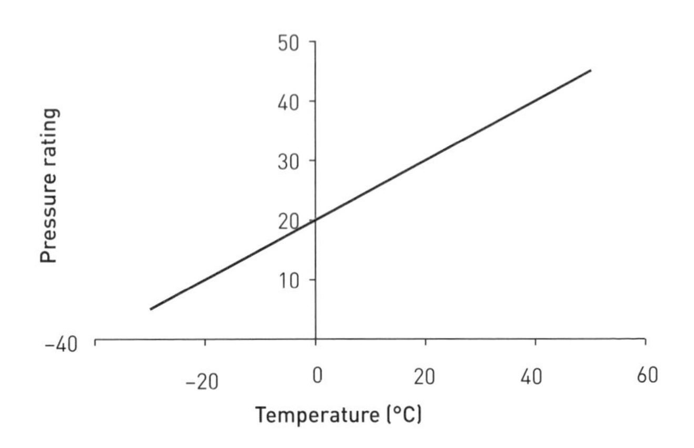

At 40°C, pressure rating is 40, at −20°C, pressure rating is 10

(b) $y = 0.5x + 20$ (c) 40 and 10 (d) 70

2 (a)

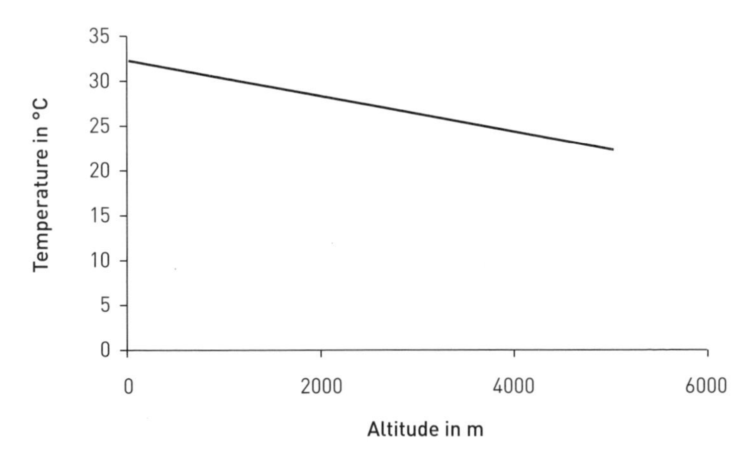

At 0m the temperature is 32°C and at 5000m it is 22°C

(b) $y = -0.002x + 32$ (c) 32°C and 22°C (d) 2°C

3 (a) $y = 30x + 100$
(b) Gradient can be taken as the cost per hour, intercept can be taken as a basic labour charge.
(c) £190

4 (a) $y = 0.29x + 25$
(b) Gradient is the cost per block, intercept can be taken as a fixed delivery charge
(c) £30.80 (d) £605

Chapter 16 An introduction to probability

Question 4 is just an extra challenge – you don't have to do it!

1 (a) $\frac{1}{6}$ (b) $\frac{1}{3}$ (c) $\frac{1}{2}$ (d) 1 (e) $\frac{5}{6}$
(f) 0 (g) $\frac{2}{3}$ (h) $\frac{2}{3}$ (i) $\frac{5}{6}$

2 (a) $\frac{1}{130}$ (b) $\frac{4}{65}$ (c) $\frac{3}{26}$ (d) $\frac{10}{13}$ (e) $\frac{1}{5}$

3 (a) $\frac{1}{8}$ (b) $\frac{3}{8}$

4 There are 16 possibilities : (H, H, H, H), (H, H, H, T), (H, H, T, H), (H, H, T, T), (H, T, H, H), (H, T, H, T), (H, T, T, H), (H, T, T, T), (T, H, H, H), (T, H, H, T), (T, H, T, H), (T, H, T, T), (T, T, H, H), (T, T, H, T), (T, T, T, H) and (T, T, T, T)
(a) $\frac{1}{16}$ (b) $\frac{3}{8}$

Chapter 17 Areas and volumes

Note that with answers involving π, if you used an approximation of 3.142 you will get slightly different answers from below – so if your answer is slightly different from mine (given to 1 decimal place) you don't need to worry.

1 (a) 45 (b) 64 (c) 35 (d) 7.5
(e) 78.5 (f) 9.8 (g) 113.1

2 (a) 84 (b) 125 (c) 226.2 (d) 179.6 (e) 41.9

3 (a) 25 (b) 24 (c) 192

Chapter 18 Logarithms

1 (a) 3 (b) 2 (c) 2 (d) 5
(e) 2 (f) 6 (g) 3 (h) 0
(i) −1 (j) −4 (k) $\frac{1}{2}$ (l) $\frac{1}{2}$

2 (a) $6 = 3 + 3$ (b) $1 = 3 - 2$ (c) $4 = 2 \times 2$

3 You cannot take the logarithm of a negative number since powers are always positive (refer back to the last question in the indices chapter), so it is impossible to create a power such that $2^? = -8$, for example.

Chapter 19 Quadratic equations

1 (a) $x = -3$ or $x = -4$ (b) $x = -2$ or $x = -6$ (c) $x = -5$ or $x = -1$
(d) $x = -6$ or $x = 2$ (e) $x = -3$ or $x = 5$ (f) $x = 2$ or $x = 4$
(g) $x = 1$ or $x = 7$ (h) $x = -4$ only (i) $x = 1$ only

2 (a) -0.232 and -1.434 (b) 3 and 0.5 (or $\frac{1}{2}$) (c) -0.228 and -8.772
(d) 0.351 and -2.851 (e) 0.768 and -0.434 (f) -0.5 (or $-\frac{1}{2}$) only

3 You end up having to take the square root of a negative number which is impossible.

Chapter 20 An introduction to trigonometry

Answers given to 3 decimal places.

1 (a) 1.414 (b) 7.616 (c) 10 (d) 13

2 (a) 140° (b) 10° (c) 2°

If two angles are 100° each then they already add up to 200° which is more than 180° so it can't possibly form a triangle.

3 sine $= \frac{5}{13}$, cosine $= \frac{12}{13}$, tangent $= \frac{5}{12}$.

4 (a) sine = 0.174, cosine = 0.985, tangent = 0.176
(b) sine = 0.5, cosine = 0.866, tangent = 0.577
(c) sine = 0.707, cosine = 0.707, tangent = 1
(d) sine = 0.866, cosine = 0.5, tangent = 1.732
(e) sine = 1, cosine = 0, tangent can't be calculated

(Note as an aside – tangent is the same as sine divided by cosine. Since in (e) the cosine is 0, this means working out 1/0 which can't be done, you can't divide by 0, and so you get an error.)

5 Have fun coming up with something – if you come up with something good, contact me via the website with your best phrases!

Chapter 21 Basic number theory

1 (a) 1, 2, 4, 8 – not prime (b) 1, 2, 5, 10 – not prime
(c) 1, 13 – prime (d) 1, 2, 3, 6, 9, 18 – not prime
(e) 1, 23 – prime (f) 1, 2, 3, 4, 6, 8, 12, 24 – not prime
(g) 1, 5 – prime (h) 1, 2, 3, 5, 6, 10, 15, 30 – not prime

2 It doesn't matter if you wrote the factors in a different order!
(a) $2 \times 2 \times 3$ (b) $2 \times 3 \times 3$ (c) 5×5 (d) $2 \times 2 \times 5 \times 5$

3 (a) 6 (b) 24 (c) 720 (d) 3628800 (e) 1

4 Primes up to 100 are 2, 3, 5, 7, 11, 13, 17, 19, 23, 29, 31, 37, 41, 43, 47, 53, 59, 61, 67, 71, 73, 79, 83, 89, 97. At the time of writing, the largest prime number known has almost 13 million digits – this may well have been broken by the time you read this!

Chapter 22 Inequalities and basic logic

1

(a) True	(b) True	(c) False	(d) False
(e) False	(f) True	(g) False	(h) False
(i) True	(j) True	(k) True	(l) False
(m) False	(n) False	(o) True	(p) False

2

(a) True	(b) False	(c) True
(d) True	(e) False	(f) True
(g) False	(h) True	(i) True
(j) False	(k) True	(l) True
(m) False	(n) False	(o) True

3

(a) 'It is very cloudy today' implies 'I can't see the sun'. Not true in reverse as for example, I may be in a room with no windows.

(b) 'The man was murdered' implies 'The man is dead'. Not true in reverse as for example, he may have died of natural causes.

(c) 'The patient has a very high temperature' implies 'The patient needs some medicine'. Not true in reverse as for example, the patient may have a normal temperature but require medication for another reason.

(d) 'The player is unfit for this match and can't play' implies 'The player won't score any goals in this match'. Not true in reverse as they may, for example, be fit and play, but not score.